Wind Energy

Wind Energy

S. C. Bhatia (Author)
BE (Chemical), MBA

Prof. Puneet Mangla (Co-Author)
B.E. (Industrial Production), M.Tech. (Engineering Systems),
Head and Associate Professor, Department of Mechanical Engineering,
Hindustan College of Science and Technology, (Mathura – UP)

Edited by

Sarvesh Devraj
B.Tech (Mechanical), UPTU
M.Tech (Renewable Energy Engineering and Management), TERI University,
(Research Associate – TERI, New Delhi)

WOODHEAD PUBLISHING INDIA PVT LTD

New Delhi

Published by Woodhead Publishing India Pvt. Ltd.
Woodhead Publishing India Pvt. Ltd.,
303, Vardaan House, 7/28, Ansari Road,
Daryaganj, New Delhi - 110002, India
www.woodheadpublishingindia.com

First published 2020, Woodhead Publishing India Pvt. Ltd.
© Woodhead Publishing India Pvt. Ltd., 2020

Woodhead Publishing India Pvt. Ltd. ISBN: 978-81-936446-2-1
Woodhead Publishing India Pvt. Ltd. e-ISBN: 978-81-936446-3-8

Typeset by Asian Enterprises, New Delhi
Printed and bound in India by Replika Press Pvt. Ltd.

Contents

Preface

Wind is a form of solar energy. Winds are caused by the uneven heating of the atmosphere by the sun, the irregularities of the earth's surface, and rotation of the earth. Wind flow patterns are modified by the earth's terrain, bodies of water and vegetative cover. This wind flow, or motion energy, when 'harvested' by modern wind turbines, can be used to generate electricity. The terms 'wind energy' or 'wind power' describe the process by which the wind is used to generate mechanical power or electricity. Wind turbines convert the kinetic energy in the wind into mechanical power. This mechanical power can be used for specific tasks (such as grinding grain or pumping water) or a generator can convert this mechanical power into electricity to power homes, businesses, schools, and the like. The urgent need to reduce the carbon footprint of human activities and the increased awareness of the consequences of climate destabilisation have rekindled interest in renewable energy sources as important elements to consider in the expansion or retrofitting of power systems. The role of wind will not only be a function of cost effectiveness and/or technology advances but also of the ability to address policy and regulatory barriers that in the past have hampered their entry into developing markets. If these barriers are successfully addressed, wind energy may contribute substantially to maintain the current, relatively low-carbon footprint of various power sectors.

This book on Wind Energy summarises various aspects of wind energy and is divided into 22 chapters.

Chapter 1 deals with wind energy: An overview. Chapter 2 focuses on global status of wind energy. The global wind energy is present today in more then 80 countries with annual instal capacity of 63 GW as on 2015. Chapter 3 is devoted to wind turbine technology. Wind turbine technology has developed rapidly in recent years and Europe is at the hub of this hightech industry. Wind turbines are becoming more powerful, with the latest turbine models having larger blade lengths which can utilise more wind and therefore produce more electricity, bringing down the cost of renewable energy generation.

Chapter 4 concentrates on small scale wind turbine technology. Small wind turbines generally have a much lower energy output than large commercial wind turbines, but their size can differ significantly. Chapter 5 focuses on planning wind projects. The development of a wind energy project is a long and complex process, involving and depending on the size of the project – the assessment of technical, economical, environmental, legal and political issues. Chapter 6 deals with wind resource assessment. Wind Resource Assessment (WRA) is the collection of technologies and analytical methods that are used to estimate how much fuel will be available for a wind power plant over the course of its useful life. Chapter 7 concentrates on offshore wind energy. The potential for offshore wind is enormous in Europe and elsewhere, but the technical challenges are also great. The capital costs are higher than onshore, the risks are greater, the project sizes are greater and the costs of mistakes are greater. Chapter 8 focuses on wind farm. A wind farm or wind park is a grouping of wind turbines in an area. Chapter 9 is devoted to major failures in wind turbines. Reliability of wind turbines is a pre-requisite to ensure the healthy growth of wind energy. Various failures such as mechanical failures, micropitting, corrosion, misalignment, etc., are discussed in detail.

Chapter 10 concentrates on improving wind turbine performance using nanomaterials. This chapter present new design scheme of light weight structure for wind turbine tower. This design scheme is based on the integration of the nano-structured materials. Chapter 11 focuses on composite material and carbon nanotubes in wind turbine blades. Developing sectors, increasing population and rising energy demand are the reasons for improvements in energy generation fields. Chapter 12 is devoted to role of lubricants in wind energy. Wind turbines require large amount of high end lubricants and also regular re-lubrication. Chapter 13 deals with wind energy powering agriculture. Different pump technologies are often flexible regarding the type of energy source powering them. Chapter 14 focuses on wind energy based desalination processes and plants. The use of Renewable Energy Sources (RES) for the operation of desalination plants is a feasible and environmentally compatible solution in areas with significant RES potential.

Chapter 15 concentrates on role of nanotechnology in non-conventional energy technology. Nanotechnology has opened the floodgate of future opportunities in enhancement of efficiency of non-conventional energy producing technologies. Chapter 16 focuses on environmental impact of wind energy. Wind power is the most economic new power plant technology, due to reduced installations costs, no fuel costs and construction time of less than one year, compared to over 10 years to construct nuclear power plants.

Chapter 17 is devoted to noise from wind energy and its control. Like any piece of equipment containing moving parts, a wind turbine emits a certain amount of mechanical noise. Chapter 18 deals with ecological aspects of wind energy. The wind farms tend to have variable effects on bird populations, which can be species-, season and/or site-specific. The impacts include collision fatalities, habitat loss and disturbance resulting in displacement. The main factors that contribute to collision fatalities are proximity to areas of high bird density or frequency of movements (migration routes, staging areas, wintering areas), bird species (some are more prone to collision or displacement than others), landscape features that concentrate bird movement and poor weather conditions. Chapter 19 is devoted to carbon footprint of wind energy. All electricity generation systems have a 'carbon footprint', that is, at some points during their construction and operation carbon dioxide (CO_2) is emitted. Electricity generated from wind energy has one of the lowest carbon footprints.

Chapter 20 concentrates on barriers to wind energy. Various key barriers such as institutional regulatory, investment and technical are discussed in detail. Chapter 21 focuses on economics of wind energy. The big driver behind the growth in wind energy investment is the falling cost of wind-produced electricity. Chapter 22 is devoted to financial and political aspects of wind energy. A financial analysis is intended to clarify whether or not the wind farm project is financially feasible for the chosen site and types of wind turbines and several other parameters.

I am thankful to Mr Sarvesh Devraj (Research associate – TERI, New Delhi) who helped me in editing the book. Appreciations are also extended to Mr Harinder Singh, Senior DTP operator, who drew and labelled the flow diagrams and worked long hours to bring the book on time. I am also thankful to the editorial team of Woodhead Publishing India Pvt. for their wholehearted cooperation in bringing out the book in time.

It may not be wrong to hold that this book on *Wind Energy* is essential reading for B. Tech./M.Tech. (Mechanical/Electrical/Chemical/Environmental/Civil Engineering). Besides students, this book will prove useful to industrialists and consultants in the respective fields.

It has been prepared with meticulous care, aiming at making the book error-free. Constructive suggestions are always welcome from users of this book.

S. C. Bhatia

Prof. Puneet Mangla

Wind energy: An overview

1.1   Introduction

Wind energy is a form of renewable energy produced through machines that use wind as their power source. Wind energy is only possible because of the sun. In recent years, wind energy has become one of the most economical renewable energy technology. Today, electricity generating wind turbines employ proven and tested technology and provide a secure and sustainable energy supply. At good, windy sites, wind energy can already successfully compete with conventional energy production. Many countries have considerable wind resources, which are still untapped.

Wind energy: Wind energy offers remarkable advantages is not used to its full potential:

1. Wind energy produces no greenhouse gases.
2. Wind power plants can make a significant contribution to the regional electricity supply and to power supply diversification.
3. A very short lead time for planning and construction is required as compared to conventional power projects.
4. Wind energy projects are flexible with regard to an increasing energy demand - single turbines can easily be added to an existing park.
5. Finally, wind energy projects can make use of local resources in terms of labour, capital and materials.

The technological development of recent years, bringing more efficient and more reliable wind turbines, is making wind power more cost-effective. In general, the specific energy costs per annual kWh decrease with the size of the turbine notwithstanding existing supply difficulties.

Many developing countries expect to see electricity demand expand rapidly in coming decades. At the same time, finite natural resources are becoming depleted and the environmental impact of energy use and energy conversion have been generally accepted as a threat to our natural habitat. Indeed these have become major issues for international policy.

Many developing countries and emerging economies have substantial unexploited wind energy potential. In many locations, generating electricity from wind energy offers a cost-effective alternative to thermal power stations. It has a lower impact on the environment and climate, reduces dependence on fossil fuel imports and increases security of energy supply. For many years now, developing countries and emerging economies have been faced with the challenge of meeting additional energy needs for their social and economic development with obsolete energy supply structures. Overcoming supply bottlenecks through the use of fossil fuels in the form of coal, oil and gas increases dependency on volatile markets and eats into valuable foreign currency reserves. At the same time there is growing pressure on emerging newly industrialised countries in particular to make a contribution to combating climate change and limit their pollutant emissions.

In the scenario of alternatives, more and more developing countries and emerging economies are placing their faith in greater use of renewable energy and are formulating specific expansion targets for a 'green energy mix'. Wind power, after having been tested for years in industrialised countries and achieving market maturity, has a prominent role to play here. In many locations excellent wind conditions promise inexpensive power generation when compared with costly imported energy sources such as diesel. Despite political will and considerable potential, however, market development in these countries has been relatively slow to take off. There is a shortage of qualified personnel to establish the foundations for the exploitation of wind energy and to develop projects on their own initiative. The absence of reliable data on wind potential combined with unattractive energy policy framework conditions deters experienced international investors, who instead focus their attention on the expanding markets in Western countries.

1.2 General characteristics of meteorology of wind

This section presents an overview of the nature of wind and the methods of assessing the wind energy potential of promising wind power sites.

1.2.1 Wind distribution

Air in motion is what we call wind. The horizontal wind speed is usually much greater than the vertical wind speed. The total energy in the atmosphere is the result of conversion of potential energy of the atmosphere into kinetic energy. The ultimate energy source is of course the Sun.

Since the Earth shape is a sphere the amount of solar energy reaching a horizontal Earth surface decreases towards the poles. Other factors affecting the energy absorbed by the Earth's surface are cloudiness, albedo of the surface (i.e., the fraction of incoming energy reflected by a surface). Absorption by aerosols and scattering are also reducing factors.

Ever since the days of sailing ships, it has been recognised that some areas of the Earth's surface have higher wind speed than others. Terms such as doldrums, horse latitudes and trade winds are well established in literature.

A very general picture of prevailing winds over the Earth surface is shown in Fig. 1.1. In some large areas or at some seasons, the actual pattern differs strongly from this idealised picture. These variations are due primarily to the irregular heating of the earth's surface in both time and position.

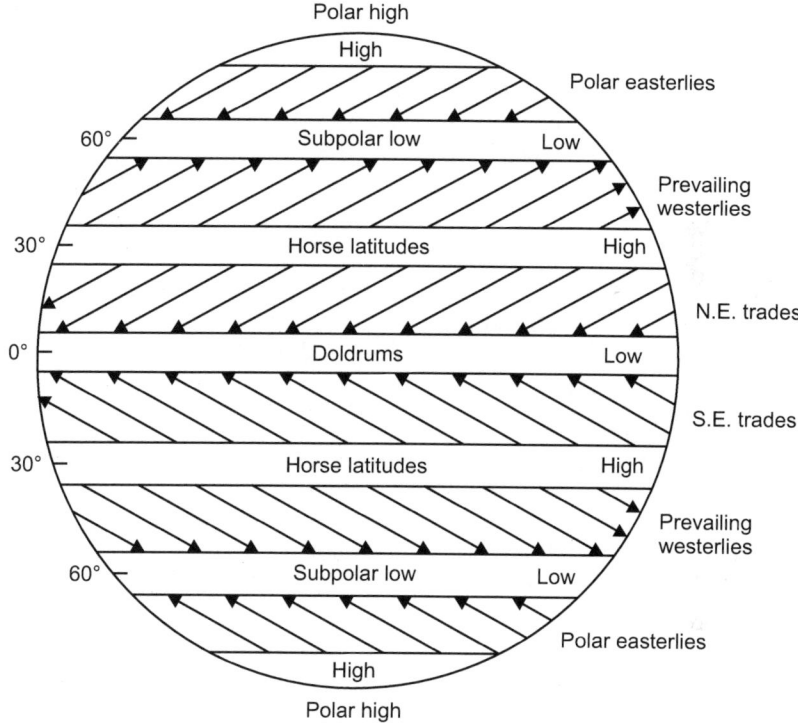

Figure 1.1: Ideal terrestrial pressure and wind systems.

The doldrums are due to a belt of low pressure, which surrounds the Earth in the equatorial zone as a result of the average overheating of the earth surface in this region. The air here is involved in strong convection flows. Late afternoon showers are common from the resulting adiabatic cooling, which is most pronounced at the time of highest daily temperature. These showers keep the humidity very high without providing much surface cooling. The atmosphere tends to be oppressive, hot and sticky, with calm winds and slick glassy seas. Unless prominent land features change the weather patterns, regions near the equator will not be very good for wind-power applications.

There are two belts of high pressure and relatively light winds, which occur symmetrically around the equator at 30° N and 30° S latitude. These are called the subtropical calms, or subtropical highs, or horse latitudes. The latter name apparently dates back to the sailing vessel days, when horses were thrown overboard from becalmed ships to lighten the load and conserve water. The high-pressure pattern is maintained by vertically descending air inside the pattern. This air is warmed adiabatically and therefore develops a low relative humidity with clear skies. The dryness of this descending air is responsible for the bulk of the world's great deserts, which are located in the horse latitudes.

There are also two belts of low pressure, which occur at some 60°S and 60°N latitude, the subpolar lows. In the southern hemisphere, this low is fairly stable and does not change much from summer to winter. This would be expected because of the global encirclement by the southern oceans at these latitudes. In the northern hemisphere, however, there are large landmasses and strong temperature differences between land and water. These cause the lows to reverse and become highs over land in the winter (the Canadian and Siberian highs). At the same time the lows over the oceans, called the Iceland Low and the Aleutian low, become especially intense and stormy low-pressure areas over the relatively warm North Atlantic and North Pacific Oceans.

Finally, the polar regions tend to be high-pressure areas more than low pressure. The intensities and locations of these highs may vary widely, with the center of the high only rarely located at the geographic pole.

The combination of these high- and low-pressure areas with the Coriolis force produces the prevailing winds shown in Fig. 1.1. The northeast and southeast trade winds are among the most constant winds on earth, at least over the oceans. This causes some islands, such as Hawaii (20°N latitude) and Puerto Rico (18°N latitude), to have excellent wind resources. The westerlies are well defined over the southern hemisphere because of lack of landmasses. Wind speeds are quite steady and strong during the year, with an average speed of 8 to 14 ms^{-1}. The wind speeds tend to increase with increasing southerly latitude, leading to the descriptive terms *roaring forties, furious fifties* and *screaming sixties*. This means that islands in these latitudes, such as New Zealand, should be prime candidates for wind-power sites.

In the northern hemisphere, the westerlies are quite variable and may be masked or completely reversed by more prominent circulation about moving low- and high-pressure areas. This is particularly true over the large landmasses.

The speed of the wind is continuously changing. Measurement of wind speed is very important to pilots, sailors, farmers, environmentalists and operators of wind energy systems. Accurate information about wind speed during some periods of time (month, year) is important in determining the best sites for wind turbines.

Wind direction, change of wind speed with height, influence of some obstacles are also important items of information. It is clear that buildings, groves and shelterbelts all have a great influence on the local flow even if the effects are damped fairly rapidly downstream. In some cases downstream effects can be felt 5–10 km after the passage of an obstacle - hills, forests, etc. Some of these effects on low level winds are due to a roughness change, e.g., from sea to land, others are due to changeable aerodynamic characteristics of different terrain features. A third category is the flow changes depending on changing thermal properties, e.g., sudden heating or cooling from below.

1.2.2   Eolian deatures

The most obvious way of recording the wind is to install appropriate instruments and collect data for a period of time. This requires both money and time, which makes it desirable to use any information that may already be available at the earth surface level, at least for preliminary investigations. The earth surface itself is shaped by persistent strong winds, with the results called eolian features or eolian landforms. These landforms are present over much of the world; they emerge on any land surface where the climate is windy. The effects are most pronounced where the climate is most severe and the winds are the strongest. An important use of eolian features will be to pinpoint the very best wind energy sites, as based on very long term data.

Sand dunes are the best-known eolian features. Dunes tend to be elongated parallel to the dominant wind flow. The wind tends to pick up the fine materials where the wind speed is higher and deposit them where the wind speed is lower. The size distribution of sand at a given site thus gives an indication of average wind speed, with the coarser sands indicating higher wind speeds.

The movement of a sand dune over a period of several years is proportional to the average wind speed. Satellite or aerial photographs easily record this movement.

Another eolian feature is the playa lake. The wind scours out a depression in the ground that fills with water after a rain. When the water evaporates, the wind will scour out any sediment in the bottom. These lakes go through a maturing process and their stage of maturity gives a relative measure of the strength of the wind. Other eolian features include sediment plumes from dry lakes and streams and wind scour, where airborne materials gouge out streaks in exposed rock surfaces.

Eolian features do not give precise estimates for the average wind speed at a given site, but can identify the best site in a given region for further study. They show that moving a few hundred meters can make a substantial difference in wind turbine output where one would normally think one spot was as good as another. We can expect to see substantial development of this measurement method over the next few decades.

1.3 Technological aspects of wind energy

Wind power is the conversion of wind energy into electricity or mechanical energy using wind turbines. The power in the wind is extracted by allowing it to blow past moving blades that exert torque on a rotor. The amount of power transferred is dependent on the rotor size and the wind speed.

Wind turbines range from small four hundred watt generators for residential use to several megawatt machines for wind farms and offshore. The small ones have direct drive generators, direct current output, aeroelastic blades, lifetime bearings and use a vane to point into the wind; while the larger ones generally have geared power trains, alternating current output, flaps and are actively pointed into the wind.

Direct drive generators and aeroelastic blades for large wind turbines are being researched and direct current generators are sometimes used.

Since wind speed is not constant, the annual energy production of a wind converter is dependent on the capacity factor. A well sited wind generator will have a capacity factor of about 35%. This compares to typical capacity factors of 90% for nuclear plants, 70% for coal plants and 30% for thermal plants.

As a general rule, wind generators are practical where the average wind speed is 4.5 m/s or greater. Usually sites are pre-selected on the basis of a wind atlas and validated with on site wind measurements.

Wind energy is plentiful, renewable, widely distributed, clean and reduces greenhouse gas emissions if used to replace fossil-fuel-derived electricity. The intermittency of wind does not create problems when using wind power at low to moderate penetration levels.

1.4 Applications and efficiency of wind energy

Most modern wind power is generated in the form of electricity by converting the rotation of turbine blades into electrical current by means of an electrical generator. In windmills (a much older technology), wind energy is used to turn mechanical machinery to do physical work, such as crushing grain or pumping water. Recently, wind energy has also been used to desalinate water.

1.4.1 Wind electric

In wind electric systems, the rotor is coupled, via., a gearing or speed control system to a generator, which produces electricity. Wind power is used in large scale wind farms for national electrical grids as well as in small individual turbines for providing electricity to rural residences or grid-isolated locations.

For small turbines the electricity generated can be used to charge batteries or used directly. Larger, more sophisticated wind energy converters are used to feed power into the grid.

Small turbines intended for battery charging have a turbine diameter of between 0.5–5 m and a power out put of 0.5–2 kW. Medium sized turbines are used in small independent grids in hybrid with a diesel or PV generator. These turbines have diameters of between 5–30 m and a power output of 10–250 kW. Large wind turbines are normally grid connected. This category includes diameters of 30–90 m and power outputs 0.5–3 MW.

1.4.2   Wind energy - water desalination

As wind energy converters supply mechanical or electrical energy, only vapour compression, reverse osmosis or electrodialysis come into consideration for wind-powered water desalination.

Wind-powered water desalination plants can be operated in island mode (with or without an additional supply of electrical energy, for example from a diesel generator set) or in grid-parallel mode. The proposal was therefore made to categorise plant configuration according to the proportion of wind energy in the desalination unit's total energy consumption. This would result in systems with a low, medium and high wind penetration rate (extent to which energy needs are met by wind). Wind energy based desalination processes and plants are discussed in detail in chapter 14.

1.4.3   Wind pumps

Wind has been harnessed to lift water for more than 2000 years, first in China and the Middle East and spreading to Europe. With wind pumps, moving air turns a 'rotor' and the rotational motion of the blades is transferred to harmonic motion of the shaft, which is used to pump water or drive other mechanical devices such as grain mills. Water from wells as deep as 200 m can be pumped to the surface by wind pumps.

In off-grid areas where there is sufficient wind (3–5 m/s) and ground water supply, wind pumps often offer a cost-effective method for domestic and community water supply, small-scale irrigation and livestock watering. To select a suitable wind pump, the following information is needed: mean wind speed, total pumping head, daily water requirement, well draw down, water quality and storage requirements.

1.5   Wind energy for development

1.5.1   Potentials

'The wind energy potential in many developing and emerging countries is substantial. In many locations, generating electricity from wind energy presents an economically viable alternative to the use of conventional fossil energy sources such as coal or diesel. In developing and emerging countries, wind turbines are an alternative to conventional power stations. In comparison to fossil-fueled power stations, wind energy can now be cost-effective in many places, as well as being non-polluting and reducing dependence on imports of fossil fuels.'

Advantages of wind can be:
1. Use of an indigenous resource without producing greenhouse gases or other pollution.
2. Wind energy contributes to the power supply diversification.
3. Wind energy projects can develop local resources in terms of labour, capital and materials.
4. Wind projects reinforce the cooperation with different donors including Germany, enhacing local capacities and technological know-how.
5. Wind projects attract new capital and can be included in the new approach of Independent Power Production (IPP).

1.5.2   Challenges of wind energy

Despite the economic and ecological advantages, so far even good wind resources in developing and emerging countries have not been used to the desirable extent. The essential reasons for this are based in the lack of knowledge in the developing and emerging countries. From the view of international wind energy companies, beside the difficulties of raising of capital and risk covering, the barriers for private investment are especially:
1. Lack of information on foreign markets.
2. Lack of knowledge of the energy-sector framework conditions and support mechanisms.
3. Insufficient wind energy legal framework (technical and economical conditions for feeding wind-generated electricity into power grids, permit procedure).
4. Lack of qualified staff, especially in the field of service/maintenance. Technicians and buyers are often unfamiliar with wind technology and in remote locations installements often break down because of a lack of servicing, spare parts, or trained manpower to administer them. In reality, wind pumps are less maintenance intensive than diesel pumps. However, the wind pump technology is 'strange' to many people and there is a need to train maintenance staff where pumps are installed.
5. Infrastructure to support the installation, commissioning and maintenance of wind generators is not developed. Users and technicians are generally unaccustomed to the technology.

6. Investment cost: Although the lifetime cost of wind is often less than diesel or petrol-powered pumps, the investment cost of purchasing a wind pump is usually higher than that of diesel pumps.

7. Wind energy does not have as consistent an output as fuel-fired power plants. Small-scale wind generators require battery storage to allow usage in periods of low or no wind. For grid connected systems, a stable grid is required to act as the storage. Wind pumps require water storage.

8. Wind generators are designed to work over a given range of wind speeds, usually 4–12 m/s. This means that the technology can only be used in areas with sufficient winds.

Global status of wind energy

2.1   Introduction

The global wind industry had another record year in 2015, with annual installations topping 63 GW. Overall, by the end of 2015, there were about 433 GW of wind power spinning around the globe, a 17% increase over the previous year and wind power supplied more new power generation than any other technology. China, the largest overall market for wind power since 2009, maintained its leadership position and installations in Asia led global markets again, with Europe in the second spot and North America closing the gap with Europe, in third place. The majority of wind installations globally were *outside* the Organisation for Economic Co-operation and Development (OECD) once again and this trend is likely to continue.

The global wind industry is present today in more than 80 countries, of which 28 countries have more than 1 GW installed, including 17 in Europe, 4 in Asia-Pacific (China, India, Japan & Australia), 3 in North America (Canada, Mexico, U.S.), 3 in Latin America (Brazil, Chile and Uruguay) and 1 in Africa (South Africa). Eight countries had more than 10 GW installed, including China (145,362 MW), the U.S. (74,471 MW), Germany (44,947 MW), India (25,088 MW), Spain (23,025 MW), U.K. (13,603 MW), Canada (11,205 MW), and France (10,358 MW). In 2016 Brazil joined the 10 GW club.

China's cumulative wind power installations (145 GW) at the end of 2015 are more than all European Union countries combined (141.6 GW). 2015 was a big year for the big markets – China, the U.S., Germany and Brazil, all of which set new records. But there is also a lot of activity in new markets around the world, Guatemala and Jordan each added their first large commercial wind farms and South Africa became the first African market to pass the 1 GW milestone. New markets are emerging across Africa, Asia and Latin America, which will provide the major growth markets for the next decade.

Asia is the world's largest regional wind market with an overall total installed capacity of 175.8 GW. In terms of annual installations China maintained its leadership position in 2015 by adding 30.8 GW of new wind power capacity to the grid, the highest annual number for any country ever.

Observers continue to be surprised by the astonishing track record for growth of the wind sector in China over the last decade. However, the Chinese wind power market may see a slowdown in 2017. Curtailment remains a major challenge in China. China's National Energy Administration and State Grid are working to solve the transmission bottlenecks and other grid issues and the situation is expected to improve over the medium term.

India's wind energy installations totalled 25088 MW at the end of 2015, keeping the Indian wind power market firmly in the top five rankings globally. Outside of China, Asia will be led by India, but new markets such as Indonesia, Vietnam, the Philippines, Pakistan and Mongolia are developing quickly.

The United States is the single largest market in terms of total installed capacity after China. The U.S. market added 4000 new turbines for a total of 8598 MW in 2015 and its total installed capacity reached 74471 MW.

Wind energy accounted for almost 31% of all new generating capacity installed in the U.S. over last 5 years. Wind energy provided more than 31% of the electricity in Iowa, 25% in South Dakota and 12% or more of the generation in a total of nine states.

The five year extension and phase out of the PTC provides the greatest degree of long term policy stability the U.S. wind industry has ever seen. This, combined with a broader range of customers and an on-going 'wind rush' driven by technological improvements is setting the stage for more years like 2015 in the U.S.

Canada's total installed wind capacity stood at 11.2 GW at the end of 2015 making it the seventh largest market globally. Canada's new wind energy projects (1506 MW) in 2015 represent over $3 billion in investment.

Mexico installed an impressive 713.6 MW of new capacity to reach a total of 4000 MW by the end of 2017. Across Europe there are now 147.7 GW installed, out of which 141.6 GW are in the EU. Wind power installed more than any other form of power generation in 2015, accounting for 44.2% of total 2015 power capacity installations. However, the overall EU installation levels mask significant volatility across Europe, 47% of all new EU installations in 2015 took place in Germany and 73% occurred in the top four markets, a similar trend was seen in 2014. This is unlike previous years when installations were less concentrated and spread across many more healthy European markets.

Germany remains the EU country with the largest installed capacity (44.9 GW), followed by Spain (23 GW), the U.K. (13.6 GW), France (10 GW) and Italy (9 GW). Sweden, Denmark, Poland and Portugal each have more than 5 GW installed. Weakened legislative frameworks, on-going economic crises and austerity measures implemented across Europe continue to hinder growth of the wind power industry. The year ahead is likely to be difficult but the broader investment shift away from fossil fuels could boost the European renewables sector. Beyond the EU, Turkey is the largest market in Europe with a cumulative installed capacity of 4694 MW at the end of 2015.

Latin America and the Caribbean has a total installed capacity of 12.2 GW. Post the Paris Agreement at COP211, the demand for clean energy bolstered by concerns for energy security and diversity of supply will promote the growth of wind power in Latin America and the Caribbean.

Brazil leads the Latin American market with installations of more than 10 GW and continues to be the most promising onshore market for wind energy in the region out to 2020.

Uruguay will generate more than 30% of its power from wind by the end of 2018 and had a total installed capacity of over 845 MW at the end of 2020, but has now surpassed the 1000 MW mark.

Chile's total installed capacity now stands at just over 1 GW, Panama added 235 MW in 2015 to reach 270 MW, and Costa Rica added 70 MW of new capacity to reach a total of 268 MW. Honduras saw its total installed capacity reach 176 MW. Guatemala for the first time added wind power to its energy mix in 2015, with a 50 MW project.

Argentina added 8 MW of new capacity in 2015 to bring its total installed capacity up to 279 MW. The Caribbean reached a total installed capacity of 250 MW across various island states.

The Pacific region saw its total installed capacity rise to just over 4.8 GW by the end of 2015. Australia, the biggest wind market in the region, brought its total installed wind capacity up to 4,187 MW.

Samoa added 550 kW of new wind power capacity in 2015. This was the first wind project in the Pacific Island nation.

The Africa and Middle East region saw 953 MW of new capacity additions in 2015, bringing cumulative capacity for the region up to 3,489 MW. Africa's wind resource is best around the coasts and in the eastern highlands, but until 2014 when the South African market took off, it was in North and East Africa that wind power has been developed at scale.

At the end of 2015, over 99% of the region's total wind installations were spread across ten countries – South Africa (1053 MW), Morocco (787 MW), Egypt (810 MW), Tunisia (245 MW), Ethiopia (324 MW), Jordan (119 MW), Iran (91 MW), Cape Verde (24 MW), Kenya (19 MW), Israel (6.25 MW) and Algeria (10 MW). New projects are expected to come online in Egypt, Ethiopia, Kenya, Morocco, Tanzania and South Africa in 2016.

2.2   Offshore wind energy

The global offshore wind industry took a big step forward in 2015, installing more than 3.4 GW across five markets globally, bringing total offshore wind capacity to over 12 GW. At the end of 2015, more than 91% (11034 MW) of all offshore wind installations were located in waters off the coast of eleven

European countries. The remaining 9% of the installed capacity is located largely in China, followed by Japan and South Korea.

Globally, the U.K. is the largest offshore wind market today and accounts for over 40% of the installed capacity, followed by Germany in the second spot with 27%. Denmark accounts for 10.5%, Belgium for almost 6%, Netherlands for 3.5% and Sweden for 1.6%. Other European markets including Finland, Ireland, Norway, Spain and Portugal make up about 0.5% of the market. The largest market outside of the European waters is China, which accounts for approximately 8.4% of the global market.

However, other countries are setting ambitious targets for offshore wind and development is starting to take off in some of these markets. Japan, South Korea and Taiwan have put actual turbines in the water. Construction is now complete on the first commercial offshore project in the U.S. and it will be commissioned before the end of 2018. The GWEC-led FOWIND consortium is developing an offshore wind roadmap for India.

Onshore wind power has become the least cost option when adding new capacity to the grid in an increasing number of markets and prices continue to fall. Also, we have recently seen record low prices in the offshore wind sector. Given the urgency to cut CO_2 emissions, clean our air and decrease reliance on imported fossil fuels, wind power's pivotal role in the world's future energy supply is assured.

The International Renewable Energy Agency's (IRENA) annual review of jobs in RE sector stated that nearly 1.1 million people were employed by the global wind power industry at the end of 2015. Driven by favourable policies and declining technology costs, rising deployment of renewables in Asian markets kept driving the regional shifts in job numbers from traditional OECD markets.

The assumption this chapter continues to make, which is verified by such studies as do exist, is that for every new megawatt of capacity installed in a country in a given year, 14 person/year of employment is created through manufacturing, component supply, wind farm development, construction, transportation, etc. While there is quite substantial regional variation, this seems to work as a global average. As production processes are optimised and thus this level will decrease to 13 person/year of employment per new megawatt installed by 2020 and to 12 person/year of employment by 2030.

In addition, 0.33 person/year of employment per MW of installed capacity are judged to be needed for operations and maintenance work at existing wind farms. Again, there will be substantial regional variations, but this seems to work as a global number.

The wind industry creates a large number of skilled, semi-skilled and unskilled jobs and this has taken on an increasing political as well as economic importance. The macro-economic effects of the development of the wind power sector as well as the renewable energy sector as a whole is increasingly a factor in political decision making on future energy choices.

2.3   Carbon dioxide reduction

Wind power's environmental benefits include the elimination of local air pollution and nearly zero water consumption. However, the greatest benefit is wind power's contribution to reduction of carbon dioxide emissions from the power sector, which is the single largest anthropogenic contributor to the global climate change problem.

Wind energy technology has an extremely good energy balance. All of the CO_2 emissions related to the manufacturing, installation, servicing and decommissioning of a turbine are generally 'paid back' after the first 3 to 9 months of operation. For the rest of its 20-year design lifetime, the turbine operates without producing any of the harmful greenhouse gases that are already disrupting life on earth. The benefit obtained from wind power in relation to CO_2 emissions depends entirely on what sort of power plant it displaces. If it displaces hydro or nuclear power, the benefit is small, but if it replaces coal or gas, then the benefit is enormous.

Emissions from fossil fuel plants range from around 500 g CO_2 /kWh up to 1200 g CO_2/kWh or more for the dirtiest fuels. On the basis of the current electricity distribution, we have calculated that 600 g CO_2/kWh is a good average number to characterise the savings generated by wind power, although the regional variations will be significant. Annual reductions in CO_2 from existing wind power plants were about 521 million tons in 2015. Under the NPS, this is expected to rise to 941 million tons annually by 2020 and up to 1987 T per year by 2030. Under the 450 Scenario, this is expected to rise to 968 million tons annually by 2020 and up to 2,293 T/annum by 2030.

The MS implies savings of over 1.17 billion tons of CO_2/annum by 2020 and more than 2.6 billion tons by 2030.

Over the long-term the AS will bring almost double the savings in CO_2 annually. In 2050 the AS is foreseen to bring about 9.2 billion tons of CO_2 emission reduction annually. In comparison the 450 Scenario forecasts annual CO_2 emission reduction of 5.6 billion tons.

2.3.1   Avoided CO_2 since 2006

In cumulative terms, the NPS has wind power saving nearly 7.2 billion tons by 2020 and more than 21 billion tons by 2030. In cumulative terms, the 450 Scenario has wind power saving nearly 7.3 billion tons by 2020 and nearly 23 billion tons by 2030.

The MS results in nearly 7.9 billion tons in cumulative savings by 2020 and 26.4 billion tons of CO_2 savings by 2030. The AS yields cumulative CO_2 savings of 8.2 billion tons by 2020 and 30.7 billion tons by 2030.

These are significant reductions across all scenarios, but the critical issue here is not just the total volume of reductions, but the speed at which these savings are achieved, as GHGs are long-lived gases and the imperative is for early emissions reductions to achieve the greatest benefit.

Wind power's scalability and its speed of deployment makes it an ideal technology to bring about the early emissions reductions which are required if we are to keep the window open for keeping global mean temperature rise to less than 2°C above pre-industrial levels.

2.4   Future of wind energy

All analyses of the global energy picture today say that the wind industry has a 'bright future'. Having experienced double digit cumulative growth for nearly twenty years, wind is unique among modern manufacturing industries, to the point where the fastest growing job in the United States is 'wind energy technician'.

Left to its own devices, market forces alone would generate steady growth for the sector, due to its low cost, speed of deployment and the stability it brings to power prices. Of course, energy markets are never left to their own devices. Public policy, tax policies, subsidies and clean air, water and climate legislation and regulation, among others, have a dramatic effect on what the power markets look like, which technologies are favoured and how the system is forced to adapt. There are two overarching questions which will have more than anything else to do with the rate and scope of wind power's expansion out to 2020, 2030 and to 2050:

1. Will humanity once and for all join forces to combat the existential threat of climate change? The Paris Agreement gives hope that this might be the case. The discourse in the current U.S. presidential election campaign does not.

2. Is there some miraculous technological breakthrough just around the corner that will transform the power sector without the need for wind, solar and other existing technologies? It doesn't seem likely, but it can never be ruled out.

2.4.1   What has changed?

Wind has become a mainstream power source: Wind provides ~4% of global electricity supply and is growing rapidly. Wind supplied more than 40% of Denmark's total power generation last year, 23% in Portugal and Ireland, ~20%

in Uruguay, 19% in Spain and 15% in Germany. The U.S. state of Iowa sourced 31% of its electricity from wind in 2015, South Dakota 25%, Kansas 24%, Oklahoma 18% and 10% in Texas and South Australia was around 40%. In 2015, wind was the largest single source (nearly 50%) of all increase in electricity generation globally.

Prices have fallen dramatically

Wind is the cheapest way to add capacity to the grid in a large number of markets, becoming the utility option of choice. Very low prices across South America and Africa and in the United States are becoming the new normal, as both the technology and the industry matures and becomes more competitive. In the U.S., the cost of wind energy has dropped by more than 65% in the past 6 years.

Integration

With the increasing penetration of wind power in a larger number of markets, differing experiences have shown that managing large penetrations of variable renewables (wind and solar) can be handled without threatening the stability of the power system and indeed, in many cases it enhances it, as the system is less vulnerable to the failure of a single large source. Increased interconnection, improved forecasting and facilities for demand management only increase possible penetration levels.

Global spread

As a look at our country by country breakdown shows, wind has moved far beyond the 'traditional' markets in North America, Europe and (more recently) China and India: Brazil, Mexico, Chile, Peru, Uruguay and Argentina, South Africa, Ethiopia, Egypt and Morocco and Iran, the Philippines, Indonesia and Vietnam are now the new markets to watch. For four out of the last five years the majority of new installations have taken place *outside* the OECD and that is expected to continue.

2.4.2   What hasn't changed?

Huge subsidies to fossil fuels and nuclear: This is many times the support given to renewables. How many times depends on what you include. While there have been some moves for subsidy reform in some countries, the basic picture hasn't changed.

International policy in the energy sector

Despite the change in rhetoric, international financial institutions are *still* spending more money supporting fossil fuels than they are renewables, with the World Bank leading the charge.

Ambiguity and instability of government policy

The steady growth and development of the renewable energy sector relies upon stable and predictable government policies. It is still the case that far too many governments talk out of both sides of their mouths and make rapid and sometimes retroactive policy changes which are very damaging to the sector. In the worlds of IEA (then Chief Economist) Executive Director Fatih Birol. 'But while variability of renewables is a challenge that energy systems can learn to adapt to, variability of policies poses a far greater risk.'

Without attempting to answer the unanswerable questions about human response to climate change and technological miracles, there are some key developments which seem very likely to occur in the wind industry over the coming decades. How fast and how far they go will depend on the broad commitment to a clean energy system globally.

Wind turbine technology

3.1 Introduction

Wind turbine technology has developed rapidly in recent years and Europe is at the hub of this hightech industry. Wind turbines are becoming more powerful, with the latest turbine models having larger blade lengths which can utilise more wind and therefore produce more electricity, bringing down the cost of renewable energy generation.

The first commercial wind farm in the U.K., built in 1991 at Delabole in Cornwall, used 400 kilowatt (kW) turbines, while the latest trials have involved turbines ten times more powerful, of four megawatts (MW) and above. The average size of an onshore wind turbine installed in 2005 was approximately 2 MW. Wind turbines have an average working life of 20–25 years, after which the turbines can be replaced with new ones or decommissioned. Old turbines can be sold in the second hand market and they also have a scrap value which can be used for any ground restoration work.

Windmills: If the mechanical energy is used directly by machinery, such as a pump or grinding stones, the machine is usually called a windmill.

Wind turbines: If the mechanical energy is then converted to electricity, the machine is called a wind generator.

Types of wind turbines: Wind turbines are classified into two general types: horizontal axis and vertical axis. A horizontal axis machine has its blades rotating on an axis parallel to the ground. A vertical axis machine has its blades rotating on an axis perpendicular to the ground. There are a number of available designs for both and each type has certain advantages and disadvantages. However, compared with the horizontal axis type, very few vertical axis machines are available commercially.

Vertical axis wind turbines: Although vertical axis wind turbines have existed for centuries, they are not as common as their horizontal counterparts. The main reason for this is that they do not take advantage of the higher wind speeds at higher elevations above the ground as well as horizontal axis turbines.

Wind turbines produce electricity by using the natural power of the wind to drive a generator. The wind is a clean and sustainable fuel source, it does not create emissions and it will never run out as it is constantly replenished by energy from the sun.

In many ways, wind turbines are the natural evolution of traditional wind mills, but now typically have three blades, which rotate around a horizontal hub at the top of a steel tower. Most wind turbines start generating electricity at wind speeds of around 3/4 meters per second (m/s), (8 miles per hour), generate maximum 'rated' power at around 15 m/s (30mph) and shut down to prevent storm damage at 25 m/s or above (50mph).

3.2 Wind turbine technology

Generating electricity from the wind is simple: Wind passes over the blades exerting a turning force. The rotating blades turn a shaft inside the nacelle, which goes into a gearbox. The gearbox increases the rotation speed for the generator, which uses magnetic fields to convert the rotational energy into electrical energy. The power output goes to a transformer, which converts the electricity from the generator at around 700 Volts (V) to the right voltage for the distribution system, typically between 11 kV and 132 kV. The regional electricity distribution networks or National Grid transmit the electricity around the country and on into homes and businesses.

3.2.1 Offshore technology

Offshore wind farms are an exciting new area for the industry, largely due to the fact that there are higher wind speeds available offshore and economies of scale allow for the installation of larger size wind turbines offshore.

Offshore wind turbine technology is based on the same principles as onshore technology. Foundations are constructed to hold the superstructure, of which there are a number of designs, but the most common is a driven pile. The top of the foundation is painted a bright colour to make it visible to ships and has an access platform to allow maintenance teams to dock. Subsea cables take the power to a transformer, (which can be either offshore or onshore) which converts the electricity to a high voltage (normally between 33 kV and 132 kv) before connecting to the grid at a substation on land.

3.2.2 Operation and maintenance

Both onshore and offshore wind turbines have instruments on top of the nacelle, an anemometer and a wind vane, which respectively measure wind speed and direction. When the wind changes direction, motors turn the nacelle and the blades along with it, around to face into the wind. The blades also 'pitch' or angle to ensure that the optimum amount of power is extracted from the wind.

All this information is recorded by computers and transmitted to a control centre, which can be many miles away. Wind turbines are not physically staffed, although each will have periodic mechanical checks, often carried out by local firms. The onboard computers also monitor the performance of each turbine component and will automatically shut the turbine down if any problems are detected, alerting an engineer that an onsite visit is required.

The amount of electricity produced from a wind turbine depends on three factors:

1. Wind speed: The power available from the wind is a function of the cube of the wind speed. Therefore if the wind blows at twice the speed, its energy content will increase eight-fold. Turbines at a site where the wind speed averages 8 m/s produce around 75–100% more electricity than those where the average wind speed is 6 m/s.

2. Wind turbine availability: This is the capability to operate when the wind is blowing, i.e., when the wind turbine is not undergoing maintenance. This is typically 98% or above for modern European machines.

3. The way wind turbines are arranged: Wind farms are laid out so that one turbine does not take the wind away from another. However other factors such as environmental considerations, visibility and grid connection requirements often take precedence over the optimum wind capture layout.

3.2.3   Grid-connected small wind turbines

Small scale wind turbines can be used in domestic, community and smaller wind energy projects and these can be either stand-alone or grid-connected systems. Stand alone systems are used to generate electricity for charging batteries to run small electrical applications, often in remote locations where it is expensive or not physically possible to connect to a mains power supply. Such examples include rural farms and island communities, with typical applications being water heating or pumping, electric livestock fencing, lighting or any kind of small electronic system needed to control or monitor remote equipment. With grid-connected turbines the output from the wind turbine is directly connected to the existing mains electricity supply. This type of system can be used both for individual wind turbines and for wind farms exporting electricity to the electricity network. A grid-connected wind turbine can be a good proposition if the consumption of electricity is high.

3.2.4   Operating characteristics of wind mills

All wind machines share certain operating characteristics, such as cut-in, rated and cut-out wind speeds.

Cut-in speed

1. Cut-in speed is the minimum wind speed at which the blades will turn and generate usable power.

2. This wind speed is typically between 10 and 16 kmph.

Rated speed

1. The rated speed is the minimum wind speed at which the wind turbine will generate its designated rated power. For example, a '10 kilowatt' wind turbine may not generate 10 kilowatts until wind speeds reach 40 kmph.
2. Rated speed for most machines is in the range of 40 to 55 kmph.

3.3 Horizontal axis wind turbines

3.3.1 Upwind turbine

The upwind turbine is a type of turbine in which the rotor faces the wind. A vast majority of wind turbines have this design. Its basic advantage is that it avoids the wind shade behind the tower. On the other hand, its basic drawback is that the rotor needs to be rather inflexible and placed at some distance from the tower. In addition, this kind of HAWT also needs a yaw mechanism to keep the rotor facing the wind.

3.3.2 Downwind turbine

The downwind turbine is a turbine in which the rotor is on the downwind side of the tower. It has the theoretical advantage that they maybe built without a yaw mechanism, considering that their rotors and nacelles have the suitable design that makes the nacelle follow the wind passively. Another advantage is that the rotor may be made more flexible. Its basic drawback, on the other hand, is the fluctuation in the wind power due to the rotor passing through the wind shade of the tower.

3.3.3 Advantages and disadvantages of (HAWT and VAWT)

Advantages

The advantages of the HAWT over the VAWT are given below:

1. Blades are to the side of the turbine's center of gravity, helping stability.
2. The turbine collects the maximum amount of wind energy by allowing the angle of attack to be remotely adjusted.
3. The ability to pitch the rotor blades in a storm so that damage is minimised.
4. The tall tower allows the access to stronger wind in sites with wind shear and placement on uneven land or in offshore locations.
5. Most HAWTs are self-starting.
6. Can be cheaper because of higher production volume.

Disadvantages

Disadvantages of the HAWT compared to the VAWT is that:

1. It has difficulties operating near the ground.
2. The tall towers and long blades are hard to transport from one place to another and they need a special installation procedure.
3. They can cause a navigation problem when placed offshore.

3.4 Vertical axis wind turbines

The vertical axis wind turbine is an old technology, dating back to almost 4000 years ago. Unlike the HAWT, the rotor of the VAWT rotates vertically around its axis instead of horizontally. Though it is not as efficient as a HAWT, it does offer benefits in low wind situations wherein HAWTs have a hard time operating. It tends to be easier and safer to build and it can be mounted close to the ground and handle turbulence better than the HAWT. Because its maximum efficiency is only 30%, it is only usually just for private use.

3.4.1 Types of vertical axis wind turbines

Darrieus turbine

The Darrieus turbine is composed of a vertical rotor and several vertically-oriented blades. A small powered motor is required to start its rotation, since it is not self-starting. When it already has enough speed, the wind passing through the airfoils generate torque and thus, the rotor is driven around by the wind. The Darrieus turbine is then powered by the lift forces produced by the airfoils. The blades allow the turbine to reach speeds that are higher than the actual speed of the wind, thus, this makes them well-suited to electricity generation when there is a turbulent wind.

Giromill turbine

The Giromill Turbine is a special type of Darrieus Wind Turbine. It uses the same principle as the Darrieus Wind Turbine to capture energy, but it uses 2–3 straight blades individually attached to the vertical axis instead of curved blades. It is also applicable to use helical blades attached around the vertical axis to minimise the pulsating torque.

Savonius turbine

The Savonius wind turbine is one of the simplest turbines. It is a drag-type device that consists of two to three scoops. Because the scoop is curved, the drag when it is moving with the wind is more than when it is moving against the wind. This differential drag is now what causes the Savonius turbine to spin. Because they are drag-type devices, this kind of turbine extracts much less than the wind power extracted by the previous types of turbine.

3.4.3   Advantages and disadvantages of VAWT

Advantages

Just like the HAWT, the VAWT also comes with a handful of advantages over the HAWT, namely:

1. Since VAWT components are placed nearer to the ground, it has an easier access to maintenance
2. Smaller cost of production, installation and transport.
3. Turbine does not need to be pointed towards the wind in order to be effective.
4. VAWTs are suitable in places like hilltops, ridge lines and passes.
5. Blades spin at a lower velocity, thus, lessening the chances of bird injury.
6. Suitable for areas with extreme weather conditions like mountains.

Disadvantages

The disadvantages of the VAWT, on the other hand are:

1. Most of them are only half as efficient as HAWTs due to the dragging force.
2. Air flow near the ground and other objects can create a turbulent flow, introducing issues of vibration.
3. VAWTs may need guy wires to hold it up (guy wires are impractical and heavy in farm areas).

3.5   Wind turbine components

Wind Turbines can be classified in two main categories based on their physical structure. Vertical axis wind turbines have a main shaft that stands perpendicular to the direction of the wind stream. Horizontal axis wind turbines have a main shaft that lies along the direction of the wind stream. Horizontal axis wind turbines are composed of the general systems (Table 3.1). The purpose of the wind turbine is to convert the kinetic energy contained in the wind to electrical energy. It is desirable to capture as much energy as possible from the wind stream and to lose as little energy as possible during the conversion process. Each of the major systems has some dependency on the other systems. Therefore, it is necessary for the turbine to have a system-wide controller which communicates with and controls various aspects of the turbine.

Table 3.1: General systems of horizontal axis wind turbines.

System	*Function*
Yaw	Track incoming wind direction
Pitch	Control blade position
Drive train	Shift torque and speed characteristic
Generator	Convert from mechanical energy to electrical energy
Power system connection	Interface generator to power grid or load systems
Supervisory controls and data acquisition (SCADA)	Monitor performance and control

The main wind turbine controller monitors the health of the entire system, as well as the health of the individual subsystems. Based on information collected from the various sensors, the main controller can set operating conditions, verify performance metrics and communicate with external parties, such as park-wide SCADA systems.

3.5.1   Pitch

Wind turbine blades provide a lift force, similar to an air-plane, which creates a torque on the main shaft. As wind passes over the blades, this force makes the shaft rotate. If there was no energy extracted from the system, via., the electrical generator and the entire system were loss less, the turbine shaft would accelerate indefinitely.

In a real system, turbulence is created around the blades as they cut through the air mass. When the speed of the shaft increases, the amount of drag force increases. Additionally, as the rotor speed increases, the blades may begin to cut into the turbulent air created by the previous blades. As the magnitude of the drag forces increases beyond that of the lift force, the blades will 'stall'. This phenomena can be taken advantage of, in that wind turbines can be designed so that they stall when the wind speed provides more power than the generator is capable of handling.

More advanced wind turbine designs allow for individual pitch control, via electric motors or hydraulic rams. Having the ability to control the pitch angle allows the designer to maximise the energy captured from the wind. Experiments can be performed to measure the performance coefficient (a measure of wind turbine efficiency) across the desired wind speed range to find a desired tip-speed ratio ($\lambda = v_{tip}/v_{wind}$). The tip-speed ratio defines the speed at which the blade cuts through the air at a given wind speed and influences the torque vs. speed curve for the wind turbine. By defining the tip-speed ratio, a specific toque vs. speed curve can be created, one that maximises the efficiency of the gearbox and generator.

3.5.2 Gearbox

A gearbox is sometimes used to increase the speed of the shaft connected to the generator. The generators characteristics will influence the choice of gear ratio. For Doubly-Fed Induction Generators (DFIGs) the gearbox ratio is generally chosen so that the desired operating wind speed range aligns with a desired generator operating speed range. For squirrel-cage induction generators, the gear ratio can be chosen so that wind speed range is beyond the synchronous speed, putting the squirrel-cage machine in the power generating region. For permanent magnet generators, the gear ratio can be chosen to increase the speed of the shaft, which has a direct influence on the operating voltage and efficiency of the generator.

Alternatively, a wind turbine with permanent magnet generator can be operated as a direct drive unit, in which the gearbox is omitted and the generator shaft is directly coupled to the main rotor shaft, this can reduce the size and weight of the nacelle, increase overall efficiency and reduce the number of moving parts and potential for failure.

3.5.3 Generator

All wind turbine generators have a generator of some kind. Although there are many different types of machines that can do the job, each offer different advantages and disadvantages. Some machines require precise control of the terminal voltages and currents, while others do not.

Some machines are capable of operating over a wide range of conditions, while others are severely limited. In general, the use of power electronics is essential to maintaining efficient operation of the generator. The generator is a shunt-field synchronous machine (basically a car alternator). This is a common machine type and is very popular among amateur wind turbine experimenters.

However, it is interesting to note that the efficiency of the turbine can be greatly affected by the way this generator is operated. Several experiments can be performed to illustrate these variations, among other topics. Also, a permanent magnet generator can be used in some experiments, it is interesting to see the tradeoffs associated with each machine type.

Woking of the generator inside the turbine

Generator consists of a rotor and stator. When the rotor is permanent magnets in this case, the stator should be stationary coils of wire and *vice versa*. When these magnets move past the coils of wire or vice versa, AC electricity is produced in the coils. It is then converted into DC electricity using a rectifier.

Parameters in wind energy production

In order to gain enough data to obtain meaningful results, three details would have to be logged in real time:
1. The output power of the turbine. (watts)
2. The rotation speed of the turbine (rev/s)
3. The wind speed (m/s)
   The analysis would require instrumentation to measure 4 parameters:
1. System voltage
2. System current
3. Turbine rotational speed
4. Wind speed

Wind turbine power output variation with steady wind speed

Figure 3.1 shows how the power output from a wind turbine varies with steady wind speed.

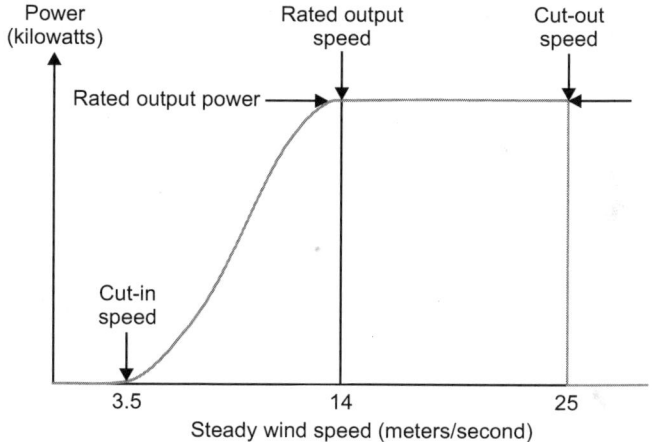

Figure 3.1: Typical wind turbine power output with steady wind speed.

Cut-in speed: At very low wind speeds, there is insufficient torque exerted by the wind on the turbine blades to make them rotate. However, as the speed increases, the wind turbine will begin to rotate and generate electrical power. The speed at which the turbine first starts to rotate and generate power is called the cut-in speed and is typically between 3 and 4 meters per second.

Rated output power and rate output wind speed: As the wind speed rises above the cut-in speed, the level of electrical ouput power rises rapidly as shown. However, typically somewhere between 12 and 17 meters per second, the power output reaches the limit that the electrical generator is capable of. This limit to the generator output is called the rated power output and the wind speed at which it is reached is called the rated output wind speed. At higher wind speeds, the design of the turbine is arranged to limit the power to this maximum level and there is no further rise in the output power. How this is done varies from design to design but typically with large turbines, it is done by adjusting the blade angles so as to keep the power at the constant level.

Cut-out speed: As the speed increases above the rate output wind speed, the forces on the turbine structure continue to rise and, at some point, there is a risk of damage to the rotor. As a result, a braking system is employed to bring the rotor to a standstill. This is called the cut-out speed and is usually around 25 meters per second.

Wind turbine efficiency or power coefficient: The coefficient of power of a wind turbine is a measurement of how efficiently the wind turbine converts the power in the wind into electricity.

The available power in a stream of wind of the same cross-sectional area as the wind turbine can easily be shown to be available power in the wind,

$$P = \frac{1}{2} \rho A v^3$$

where, v is wind speed in meters per second, ρ is density in kilograms per cubic meter and A is the swept area of the trubine ($A = \pi r^2$), r is rotor radius in meters then the available power, P is in watts.

The efficiency, μ, or as it is more commonly called, the power coefficient, cp, of the wind turbine is simply defined as the actual power delivered divided by the available power.

Coefficient of power = (Power produced by the wind turbine/Total power available in the wind)

The Betz limit on wind turbine efficiency: There is a theoretical limit on the amount of power that can be extracted by a wind turbine from an airstream. It is called the Betz limit.

Betz limit = 16/27

3.5.4 Power system interconnection

Wind turbines can be operated as part of an existing power distribution network, or in a stand alone 'island' power system. In either case, it is important that the power generated by the turbine is made available for use by the system. This generally requires the use of power electronic controllers, transformers, filters and additional protective devices. Furthermore, the wind turbine must be capable of limiting its power output and must have the capability of withstanding various fault conditions. For instance, when connecting a wind turbine to an existing power distribution network, it is important that the terminal voltage and frequency match that of the power system. When used in a stand alone setting, the wind turbine may play a large role in the nature of the power system dynamics and additional requirements or control capabilities may be imposed to maintain the integrity of the system.

3.5.5   Supervisory control and data acquisition

Supervisory Control and Data Acquisition (SCADA) systems are an important item to consider. SCADA systems can collect information from wind turbines, substations, loads and system operators and can control turbine set-points to maintain reliable operation. When power generation signals are provided by a system operator, such as Mid-Continental Independent Service Operator (MISO), the SCADA system adjusts the set-points of individual turbines and can shut down turbines in the case of excessive energy production.

SCADA systems also provide the operator with visual information regarding turbine status and health. Interfaces are usually provided to see system details and to provide the ability to remotely control the wind turbine.

One single wind turbine may not be able to produce desired level of electricity, hence, numbers of wind turbines are connected together to obtain desire output. This assembly of wind turbines together is called wind farm. We must choose the place for constructing wind farm where the wind speed is sufficient to move the blade of turbine. When the wind flows through the blades of turbine, the turbine rotates for running a generator to produce electricity. This electricity flows down through the cable attached to turbine tower. This cable is also interconnected with cables from other wind turbines in the wind farm.

3.6   How does wind turbine work?

There is a light weight turbine of large diameter attached to the top of a supporting tower of sufficient height. When wind strikes on the turbine blades, the turbine rotates due to their typical design and alignment. The shaft of the turbine is coupled with an electrical generator. The output of the generator is collected through electric power cable.

3.6.1   Working of wind turbine

When the wind strikes the rotor blades, blades start to rotating. Rotor is directly connected to high speed gearbox. Gearbox converts the rotor rotation into high speed which rotates the electrical generator. An exciter is needed to give the required excitation to the coil so that it can generate required voltage. The exciter current is controlled by a turbine controller which senses the wind speed based on that it calculate the power what we can achieve at that particular wind speed. Then output voltage of electrical generator is given to a rectifier and rectifier output is given to line converter unit to stabilise the output AC that is feed to the grid by a high voltage transformer. An extra units is used to give the power to internal auxiliaries of wind turbine (like motor, battery, etc.), this is called Internal Supply unit (ISU). ISU can take the power from grid as well as from wind. Chopper is used to dissipate extra energy from the RU for safety purpose. Figure 3.2 shows internal block diagram (ISU) of wind turbine.

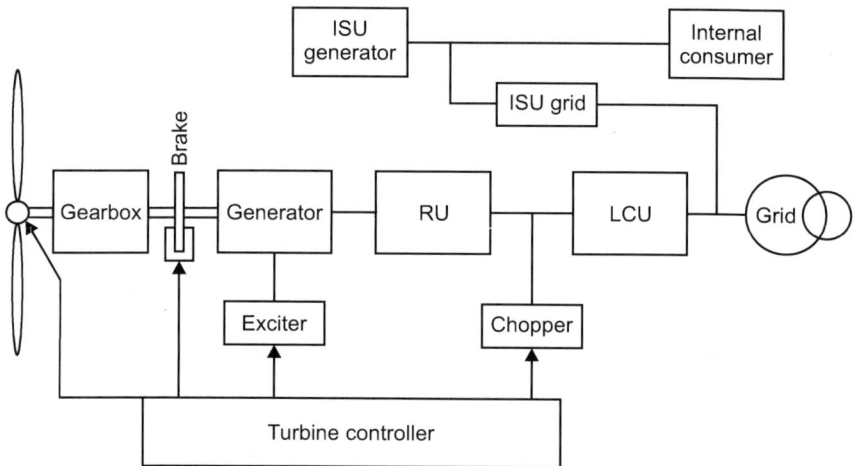

Figure 3.2: Internal block diagram (ISU) of wind turbine.

3.6.2   Major parts of wind turbine

Tower is very crucial part of wind turbine that supports all the other parts. It not only supports the turbine but raises the turbine to sufficient height so that its blades tips would be at safe height during rotation. Not only that, the height of the tower is so maintained that it can get sufficiently strong wind. The height of tower ultimately depends on the power capacity of wind turbines. The tower of the turbines in commercial wind power plants usually ranges from 40 meter to 100 meter. These towers may be either tubular steel towers, lattice towers, or concrete towers. The tubular steel tower are used for very large wind turbine. These are normally manufactured in section of 30 to 40 meter length. Each section has flanges with holes.

Such sections are fitted together by nut bolts at site to form a complete tower. The complete tower is slight conical shape shape to provide better mechanical stability. Lattice tower is assembled by different members of steel or GI angles or tubes. The all typical members are bolted or welded together to form a complete tower of desire height. The cost of these towers are much less than that of steel tubular tower but aesthetic look is not as good as steel tubular tower. Although, transportation, assembling and maintenance are quite easy but still use of lattice tower is avoided in modern wind turbine plant due to its aesthetic look. There are another type of tower used for small wind turbine and this is guyed pole tower.

This is a single vertical pole supported by guy wired from different sides. Because of numbers of guy wires, it is difficult to access the footing area of the tower. Because of that this type of tower is avoided in agricultural field. There is another type of wind turbine tower used for small plant and this is hybrid type tower. This is also a guyed type tower but only difference is that instead of using a single pole in the middle it uses a thin and tall lattice type tower. This is hybrid of both lattice type and guyed type tower.

Small scale wind turbine technology

4.1   Introduction

Small wind turbines generally have a much lower energy output than large commercial wind turbines, but their size can differ significantly. So called Micro wind turbines may be as small as a fifty watt generator and generate only about 300 kWh per year. They are used for boats, caravans, miniature refrigeration unit, but also for fence-charging and other low-power uses. In comparison to that household-size turbines reach diameters of 9-meter, can have a rated power of 20 kW and produce about 20000 kWh per year for homes, farms, ranches and small businesses. The biggest turbines described as small-scale wind turbines have a rated power of 50 kW. Small units often have direct drive generators, direct current output, aeroelastic blades, lifetime bearings and use a vane to point into the wind. Larger, more costly turbines generally have geared power trains, alternating current output, flaps and are actively pointed into the wind.

4.2   Small wind turbine technology

Nearly all small wind turbines today are upwind, horizontal-axis turbines, which means the rotor is spinning in front of the tower. There are some turbines using two blades, while the majority of the recent turbines comes supplied with a three-blade rotor, which in general terms makes the turbine run more smoothly and last a longer time. The prevalent blade materials are composite materials as fiberglass, while only a few products in this class still use wood. Instead of the yaw motors of the big wind turbines, small wind turbines often use tail vanes to point the rotor to the wind.

Micro and Mini wind turbines use generators based on permanent-magnet alternators. The magnets in the generator are conventionelly tied on the rotating shaft driven by the rotor, but their are also several wind turbines with the magnets attached in a case which rotates around the stationary part of generator. This inside-out design has two advantages.

The blades can be bolted directly to the case containing the magnets and the magnets are pressed against the case wall by the centrifugal force. Contrary to this, conventionelly attached magnets on the rotating shaft of a generator have to be retained by sophisticated means.

In household-sized generators, besides permanent-magnet alternators conventional wound-field and induction alternators are used. Most small wind turbines generate a three-phase AC which can be rectified by a controller for battery charging applications. There are turbines with built-in controllers and also products with an external controlling entity. Because of the sometimes demanding environmental conditions, the robustness of a turbine is a very important parameter, which can be estimated only very roughly. Experience has shown, that the weight of the turbine in relation to the area swept by the rotor can be used as a criterium. For example a turbine with a relative mass of 10 kg/m^2 should be more robust than a turbine with 5 kg/m^2.

To prevent damages caused by very high winds, every wind turbine needs a means for overspeed control. The preferred mechanism used by producers is furling or folding the turbine across a hinge so that the rotor swings towards the tail vane. In case of a passive furling mechanism, the thrust of very high winds overcomes the restraining force which kept the rotor towards the wind. The threshold value (wind speed) for this mechanism is caused by the design of the hinge, which connects the body of the turbine and the tail vane. Furling mechanisms are very common for micro and mini wind turbines, while many household-sized turbines pitch the rotor-blades for overspeed control and a few also use a combination of pitching and furling.

4.3 Classes of small wind turbines

4.3.1 Comparing small wind turbines

Wind turbines are usually compared by their rated power in W, kW or MW. It is important to look at these values with care, because there is no standard in rating the output of small wind turbine output. The most significant differences are revealed in the diverging wind speeds related to the rated output for a wind turbine. The power contained in a wind with 12.5 m/s is almost three times higher than at a wind speed of 9 m/s. Thus the rated power of a wind turbine given for a wind speed of 12.5 m/s is three times higher than the value given for a wind speed of 9 m/s. The same machine can be labelled with a very different rated power only depending on the wind speed which is used as basic value. It must be mentioned, that only at the windiest sites of the world turbines will operate for a significant time span at a wind speed of 12.5 m/s. At most sites such high winds are very rare. Looking at the rated power of a wind turbine one must compare the wind speed used in the rating procedure with the expected wind speeds at the site the turbine will be installed.

As an alternative the energy (in kWh) that is generated by a small wind turbine at a site with a given average wind speed can be used to compare the turbines. Finally the combination of rotor diameter and energy generated per year (at a reference wind speed) is a roughly reliable indicator for comparing small wind turbines.

4.3.2 Micro and mini wind turbines

Micro wind turbines

These wind turbines have a very small rotor diameter of around 1 m or less and generate about 300 kWh per year at sites with an average wind speed of 5.5 m/s. They are typically used for low-power uses in remote areas (e.g., fence-charging, basic lighting, electricity for sailboats).

Mini wind turbines

Mini wind turbines typically have rotor diameters of 1.5–2.6 m an generate 1000–2000 kWh per year at sites with 5.5 m/s.

4.3.3 Household-size wind turbines

This term summarises a much broader field of wind turbines depending on the very different size of 'households' and the related applications. Household-size turbines are suitable for the supply of homes, but also for farms, ranches and even for small businesses and telecommunications. Thus in this class rotor diameters of 2.7–9 m can be found and the generated energy per year ranges from 2000–20000 kWh for sites with 5.5 m/s.

4.4 Siting and safety

Finding the right place for the wind turbine can be the most challenging part of the installation. The most important parameters for this choice are given by the terrain surrounding the site of the wind turbine. Tower height and distance from buildings or other obstacles have to be considered carefully, while the related costs often are a limiting factor. For siting a small wind turbine there is an old rule of thumb called the 30-foot (10-meter) rule. For good performance the wind turbine should be installed at least 10 m above all obstacles within 100 m. This is a minimum to avoid placing the wind turbine in the disturbed wind flow around trees or buildings. Thus in a perfect surrounding of a grass land without any obstacles, a tower should be at least 10 m high.

To safe costs it is essential to use any given advantages of the landscape. If there is a hill near the building or village which should be supplied with electricity, the best place for the new wind turbine will be on the top of the hill. Choosing the distance between the tower and the supplied building, the cost of connecting and the burial of the cable have to be considered. Installing a wind turbine directly on a roof is problematic for three reasons. The turbulence in the wind stream caused by the building is very high and the performance will be significantly decreased. Secondly, small wind turbines can be very noisy especially in high winds and a last very important reason is the risk for the people living in the house. Modern small wind turbines are durable and reliable but nevertheless it must be avoided by all means to work or live beneath a machine with parts moving as fast as the blades of a small wind turbine.

4.5   Installation of small wind turbines

For micro and mini wind turbines tilt-up towers are the easiest and cheapest carrying constructions. A tube with a diameter dependent on the size and weight of the wind turbine should carry is tilted over a hinge fixed in the ground until the tower is raised vertically. A so-called gin pole is used to raise the tower over the hinge. A gin pole is a second shorter mast at a right angle to the tower, which acts as a lever to reduce the lifting loads. Bigger household-size wind turbines require heavier lattice towers which can only be erected by a crane. The following section focuses on raising a tilt-up tower, which needs only a few persons and relatively little experience and skills.

Common tower tubes consist of several pieces with lengths of a few meters, some smaller towers have separations of only two meters length allowing a comfortable delivery by standard parcel services. Tilt-up towers are anchored by guy cables, whose number and diameter depends on the height of the used tower. The anchors for the guy cables could be screwed into the ground, if the soil is medium-dense. Using these screw-anchors simplifies the installation of a tower a lot, but the stability of their placement must be proven with great care. For heavier wind turbines, bad ground conditions (very low or very high density) or very high wind areas conventional concrete anchors or power-driven screw anchors are required.

The power conductors carrying the generated power to the ground are threaded through the inside of the tower tube in most cases. Their attachment is a very important but often overlooked aspect of the installation. If the conducting cables are connected to the leads of the wind turbine only, the leads are likely to be pulled out and damaged by the weight of the conductor. For this reason a strain relief has to be used to support the weight of the cable. A strain relief is a fine-meshed wire net which can be put around the cable. The upper side of the strain relief is fixed at a point at the top of the tower. Additionally the connection between the wind turbine leads and the cable should be done by an in-line vibration proof compression connector.

4.5.1   Tools for the installation of small wind turbines

For the erection of a tilt-up tower with a wind turbine either an electrical winch or a type of hoist is necessary. A relatively light-weight, low-cost and very effective tool working like a hoist in principle is the so called griphoist.

In contrast to a winch a griphoist does not furl the cable, wire or rope, but it pulls it directly through a hoist inside the tool. The griphoist is used manually by a lever and needs no supply of electricity. Combined with the low weight it is an ideal tool for raising tilt-up towers especially in remote areas, where an electrical winch and the necessary battery are not available. Besides practitioners prefer these tools because of the full and direct control the operator has over the raising process. The force needed by the operator to move the lever is a good indicator for the status of the raising procedure. In this way the operator is able to react on unusual tensions on the cable and can reduce the lift, when the tower is near to vertical.

4.5.2　Raising procedure

If team has little experience in erecting a tower, it will be favourable to practise the whole work flow before you raise the tower with the much greater weight of the turbine. In this way site-specific difficulties can be identified without the risk of damaging the most expensive and important part. the wind turbine.

As a first step the gin pole of the tilt-up tower must be raised connecting a cable at the end of the gin pole (with a shackle) and pulling it to a vertical position by the griphoist. Afterwards the anchor for the griphoist must be placed exactly at the position the end of the gin pole will reach, when the tower is upright. The reason is the common construction of the tower and the gin pole. Because the gin pole consists of several pieces of tube, the force executed by the griphoist has to pull downwards and must not pull from a position which is further from the tower base than the length of the gin pole. Otherwise the tube pieces could come apart under the high tension during the lifting process endagering the tower, turbine and especially the working people.

When the griphoist is anchored carefully and the lifting cable is connected to the end of the gin pole, the tower can be erected. While one person is operating the griphoist, for minimum one additional person is needed keeping a guy cable tensed and guiding the lift while preventing any jerks during the process. It is very important to keep this guiding person well outside the fall zone at right angle to the tower. Depending on the size of the mast, the raising procedure can take several hours, because with every lever-movement at the griphoist, a few inches of cable are pulled through the hoisting mechanism. Lifting weight is greatest when the tower is near the ground and decreases with every degree the tower is lifted towards vertical position. After practising the raising procedures a couple of times the wind turbine can be attached and the procedure can be repeated. When the tower reached a vertical position the guying cables have to be connected and tightened to the anchors. Again it is important to raise the tension on the guy cables step by step to prevent the thin tube material of the tower from buckling.

4.6 Common applications of small wind turbines

4.6.1 Village power: Potable water

Small mechanical wind turbines have been used to drive pumps for potable water for a very long time. Today small electric wind turbines are an efficient alternative which can be used to supply people and livestock with underground water from a well. Creating a village water tap eliminates the need to carry water from distant sources and using underground water generally avoids common health problems. The size and capacity of the needed generator is proportional to population served and pumping height. For example a turbine with a capacity of 1 kW can supply electricity approximately to 200 people.

4.6.2 Village power: Productive uses

Examples for productive uses in rural areas are irrigation, agroprocessing or ice-making. Small wind systems can be an excellent foundation for electrification, which at the same time increases income and chances for cost recovery of the system. The previously mentioned uses require more energy than pumping for drinking water. A turbine with a 1 kW generator can approximately support the work of 10 people in this case.

4.6.3 Village power: Pre-electrification

Pre-electrification means installing a small source of electricity for basic needs like lighting. Thus the costs for candles, kerosene or dry-cell batteries can be saved. The systems often work with micro wind turbines (25–120 Watts per household), have no grid connection, but generate a direct current that can be used for charging batteries. High efficiency fluorescent bulbs are often used to make the most advantage from the small lighting systems.

Planning wind projects

5.1   Introduction

The development of a wind energy project is a long and complex process, involving and depending on the size of the project – the assessment of technical, economical, environmental, legal and political issues.

Although a lot of expertise is needed to conduct the necessary assessments and to evolve an appropriate project layout, as a central task the developer has to bear in mind and control the conduction of the basic steps of wind energy development. This chapter describes the activities to be done, but does not give step-by-step solutions, because the framework conditions of projects vary case by case. Summary of the various aspects that play a role in the development of wind energy projects is given below.

5.2   Pilot study/initial site selection

The initial site selection is the first phase in the development of any wind energy project. In this phase appropriate sites should be identified and their wind potentials should be estimated. By identifying environmental, technical, commercial and political constraints of the sites the project developer can decide whether a more extensive feasibility study should be conducted. As a starting point many developers visit the possible project sites, gathering first impressions about topography and infrastructure (roads, dwellings, grid-connection). As a central task in this phase, available environmental and technical data must be collected.

5.2.1   Site selection

The 'desk-based' studies for site selection are conducted to decide whether site characteristics fulfill crucial technical criteria for the successful development of a wind project:

1. The developer will usually identify sites with sufficient potential for a suitable wind resource by using a combination of maps of the area, results of computer modelling, meteorological offices (airports, harbours, farming), or data from university departments dealing with wind energy. Promising values are average wind speeds above 6 m/s. Characteristics of other wind parks in the area have to be investigated.

2. The local road network must be suitable to provide access for large transportation vessels. An initial investigation will give a first idea of the necessary extensions for the wind project.
3. Site ownership must be considered.
4. Grid connection must be available in an appropriate distance to keep connection costs low.

5.2.2 Considerations about legal aspects for the site selection

The initial questions related to legal aspects of the project development is concern with the environmental status of the area. To gain information about the feasibility of a wind project within the constraints of environmental protection at potential sites, the developer has to consult local planning authorities for reports, maps and studies about the environmental status of the proposed site. This status includes information about ecological designations concerning areas or protected species. Due to the relatively long time period necessary for developing a wind project, also changes in the environmental policy and planned but not yet realised protected areas have to be taken into account. Besides this 'purely environmental' issues the considerations have to cover the following aspects of the site:

1. Visual aspects: The visibility of the proposed project from important public viewpoints has to be checked.
2. The distance to domestic dwellings should be sufficient to avoid disturbance of the inhabitants by noise, shadow flicker, visual domination or reflected light.
3. Recreational uses: Development plans of local planning authorities have to be checked for sites and areas dedicated to recreational use.
4. Civil and military airports: Local airport authority have to be consulted.
5. Proximity of archaeological/historical heritage sites could be constraints for the development of a wind project.
6. Telecommunications: As wind turbines can affect microwave connections, TV, radar or radio transmissions adversely, position of masts and other infrastructure must be investigated.
7. Restricted areas: Military installations but also telecommunications installations can be reasons for restrictions for the development of wind projects in the surrounding area.

Dialogue and consultation

At this stage of a project, dialogue with local authorities has to be opened to gather information and to define the major issues related to the planning process which have to be discussed in more detail in the following planning phases. The officers of the local authorities may also recommend other consultants with experiences in wind energy development in the area. The focus of the consultation for site selection is gathering and distributing information.

1. Evaluation of the quality of wind data collected during the pilot study is based on the outcome of this evaluation the developer has to decide, whether own wind measurements are necessary for reliability of the estimation of the expected profits of the project.

2. Common and established methods for site evaluation should be applied.

3. Wind speed distribution is measured for an appropriate period of time (minimum one year, in case no other reliable sources are available). Measurement for large wind projects requires installation of masts of heights about 60 m. For the placement of these masts an own planning application is often necessary. This process extends the project development significantly. As small differences in the prediction of wind speed distribution causes large changes in potential wind energy yield, a measurement campaign is essential to create sufficient reliability for potential investors.

5.2.3   Building ground

The soil conditions are assessed to define basement types: Flat basements can be used in case of solid ground conditions but in for soft soil pile basements may be necessary. Each proposed placement site of a turbine tower needs an own survey. A very important part of the site evaluation is the development of a detailed description of possible environmental impacts of the wind project.

In European countries planning authorities often request an environmental statement, containing a number of assessments concerning the following environmental issues.

1. Justification of the site selection.

2. Visual and landscape assessment - description of the landscape and the possible impact of the project.

3. Assessment of the noise generated by the wind turbines - conclusions for the chosen distance to domestic dwellings.

4. Ecological assessment.

5. Archaelogical assessment.

6. Hydrological assessment.

7. Interference with telecommunications systems.

8. Aircraft safety.

9. Safety assessment - technical description of the integrity of the chosen turbines for the site.
10. Traffic management and construction - like the impact of the wind turbines the construction of the access roads has to be discussed with local authorities.
11. Electrical connection.
12. Effects on the local economy.
13. Tourism and recreational effects.
14. Global environmental effects: By the estimation of the energy yield of a project, the expected avoided emissions can be estimated.
15. Decommissioning: The environmental statement must contain considerations about removal of the wind turbines and the related restoration processes of the landscape after the project period.

Grid

The local electricity distribution system has to be examined by available plans and consultation of the local electricity company. The dialogue with this company reveals whether an electrical connection to the sites under consideration is technically and commercially feasible, because the company can give an indication of the likely costs of the wind project connection to the grid.

5.3   Planning wind projects

5.3.1   Planning

1. Wind park layout is planned based on wind speed distribution, surface parameters and environmental conditions. Type and rating of wind turbines are selected and the layout is optimised by computer tools concerning the expected output. Besides output, installation of connection lines and (possible) transformer station as well as construction of roads for installation and service of the wind park are essential criteria for the project layout.
2. The outcome of soil surveys indicates, whether the optimal placements determined by modelling can be realised.

WEC/Grid

1. The choice of number, type and rated power is influenced by available grid capacity at the site. The cost of building a transformer station or long grid connections have to be evaluated as initial costs in comparison to expected profits.

2. The calculation is a complex optimisation depending on costs of the turbines, cost of grid connection and profits depending on expected wind conditions.

Financing

All costs related to the proposed layout have to be evaluated including wind turbines, basements, grid-connection, road construction, planning and monitoring of the construction process, advice about legal and tax conditions, cost of planning, project approval and environmental studies. Examination of possible support mechanisms (political framework conditions like feed-in tariffs or other funding mechanisms). Clean Development Mechanism (CDM) baseline studies can be prepared in case access to CDM is an option.

Type of company

The legal body for the operation of the wind project must be chosen. The choice depends partly on the number of people involved in the project. In general a larger number of people (e.g., a community as a whole) is a great advantage for the development of the wind project. A limited liability, a co-operative, or a joint-venture with a local energy company are feasible alternatives.

Economic feasibility

All relevant data of the wind project is combined to evaluate the economical feasibility of the wind project:

1. Wind speed distribution and expected energy yield.
2. Expected feed-in returns.
3. Initial investment costs for the proposed project layout.
4. Annual costs for financing.
5. Annual costs for maintainance.

The expected annual profit is calculated by comparison of costs and returns of the project. Additionally the overall profit is calculated by summing up the annual profits over project lifetime.

5.3.2   Building application and environmental impact review

After all detailed technical, economical and environmental assessments are completed, the project developer may submit a planning application to the local planning authorities.

This application is normally submitted in combination with the environmental statement describing the outcome of all assessments conducted by the product developer.

With the submission of the application, the project developer starts a phase of discussion and co-operation:

1. The planning application has to be made available to the public to allow inspection and response.

2. Based on the environmental statements, the planning authorities may implement additional regulations concerning noise control, traffic safety, decommissioning and restoration, interferences with telecommunications, ecological impacts as well as the design and colour of the turbines.

3. If the project developer and the planning authorities close discussion with a suitable outcome and the requested changes in the project development are implemented, the planning application is approved and the project permission is granted.

Wind resource assessment

6.1   Introduction

For any power plant to generate electricity, it needs fuel. For a wind power plant, that fuel is the wind. Wind Resource Assessment (WRA) is the collection of technologies and analytical methods that are used to estimate how much fuel will be available for a wind power plant over the course of its useful life. This is the single most important piece of information for determining how much energy the plant will produce and ultimately how much money it will earn for its owners. The resulting estimates of energy production are most often used by project developers to secure funding from investors to build a project. For a wind project to be successful, accurate wind resource assessment is essential. Given the cost and effort associated with conducting a wind resource assessment campaign, this type of undertaking is most often only associated with utility-scale projects, consisting of multiple wind turbines with nominal generating capacities upwards of 1.0 megawatts each.

The technology for measuring wind speeds has been available for centuries. The cup anemometer - the most commonly used type for wind resource assessment - was developed in the mid-19th century and its basic design (three or four cups attached to a vertical, rotating axis) has scarcely changed since.

Yet an accurate estimate of the energy production of a large wind project depends on much more than being able to measure the wind speed at a particular time and place. The requirement is to characterise atmospheric conditions at the wind project site over a very wide range of spatial and temporal scales - from meters to kilometers and from seconds to years. This entails a blend of techniques from the mundane to the sophisticated, honed through years of sometimes onerous experience into a rigorous process.

This chapter discusses – where does the wind come from? What are its key characteristics? And how is it converted to electricity in a wind power plant?

Where do winds come from: The simple answer to this question is that the air moves in response to pressure differences, or gradients, between different parts of the earth's surface. An air mass tends to move towards a zone of low pressure and away from a zone of high pressure. Left alone, the resulting wind would eventually equalise the pressure difference and it would die away.

The reason air pressure gradients never completely disappear is because they are continually being powered by uneven solar heating of the earth's surface.

When the surface heats up, the air above it expands and rises and the pressure drops. Wind energy enjoys a unique place in climate change mitigation. It is the most mature and the most cost-effective solution to reducing global greenhouse gas (GHG) emissions from the electricity sector, which is the largest contributor of GHG emissions.

The WRA, specifically wind measurement, is in most cases the longest lead time activity in a wind project. It is also the most pivotal activity because it determines the bankability of the wind project. The developing member countries face unique technical and financial challenges during the WRA process, which hamper wind project developers' ability to obtain financing. Although conditions in each country are different, this chapter highlights the common challenges and proposes solutions, with the goal of accelerating the development phase of a wind energy project. The underlying message of the guidelines in this chapter is that execution of a WRA—wind measurement, wind flow modelling and estimation of losses and uncertainty—requires rigor and experience.

Wind power is the fastest-growing renewable energy worldwide. In countries with viable wind resources, deployment of wind power features prominently in the renewable energy road map. A pivotal activity during the development phase of a wind project is wind resource assessment (WRA).

WRA quantifies the wind resource, therefore, it is a key activity not only to determine the financial feasibility of a specific wind project, but also to map the wind resources for a region or a country.

In developing wind energy markets, it is common to encounter bankable WRAs that do not meet international standards with respect to (i) on-site wind measurement with tall towers and high-quality instruments for a period of at least 1 year, (ii) auditable data acquisition, processing and archiving, (iii) rigorous modelling of wind flow based on a linear model for simple terrain and a computational fluid dynamics model for complex terrain and (iv) a credible estimate of losses and uncertainties.

A poorly executed WRA that does not meet international standards can significantly increase the lead time of the wind project, for example, from 2 years to over 5 years. In such cases, in addition to the delay, the developer is forced to invest additional capital. A strong and committed developer may continue with more investment, but large fraction of financially weak developers choose to abandon projects, even though these projects hold development licenses and land concessions.

Recommended strategies to assist developers and increase chances of success for wind projects in the development phase include the following:

1. Offer grants or low-interest loans to qualified developers with strict conditions that pertain to use of best practices for bankable WRAs.

2. Create licensing guidelines that clearly state conditions on the quality of WRAs prior to issuing a generation license.
3. Develop capacity of developers through knowledge transfer and training.
4. Develop long-term wind measurement campaign to collect high-quality long-term wind data.

All these strategies lead to a higher probability that the projects under development are financed and subsequently built. The output of a bankable WRA is average annual energy production (AEP), net AEP after losses and uncertainty associated with net AEP. This information is used in financial analysis of the project, with uncertainty playing a prominent role.

6.2 Wind resource assessments

Wind energy is widely recognised as one of the cheapest forms of clean and renewable energy. In fact, in several countries, wind energy has achieved cost parity with fossil fuel-based sources of electricity generation for new electricity generation plants. The input raw material (feedstock) for a wind power plant (WPP) is wind, therefore, there is no pollution or environmental degradation due to mining and/or transporting of raw materials like coal, crude oil, or natural gas. The output is clean electricity with absolutely no pollutants—no carbon dioxide, sulphur oxides, nitrogen oxides, or mercury compounds— and the WPP does not use water. All these favourable characteristics of WPPs provide significant tangible benefits in terms of improved health and conservation of natural resources.

WPPs use wind as the only resource and their output therefore depends on the wind resource present on-site. Wind resource assessment (WRA) is the discipline of quantifying the wind resource.

6.3 Wind energy project life cycle

The seven activities in the life cycle of a wind project are shown in Fig. 6.1. The duration of each activity and a description of the activities are given in Table 6.1. WRA is a pivotal activity in the life cycle of a wind project because it governs the financial viability of the project. It involves wind measurement for at least 1 year, modelling of wind flow, computation of Annual Energy Production (AEP) and estimation of losses and uncertainty associated with AEP. In new wind markets, this activity may stretch to over 5 years, as financiers demand longer measurements, with taller meteorological towers (or met-masts) and with higher-quality instruments. The reason is that in new wind markets, there is sparse or no high-quality wind data that can be used to verify the measured data and to verify the wind resource model's extrapolations into the future, to the entire wind-farm area and to hub height.

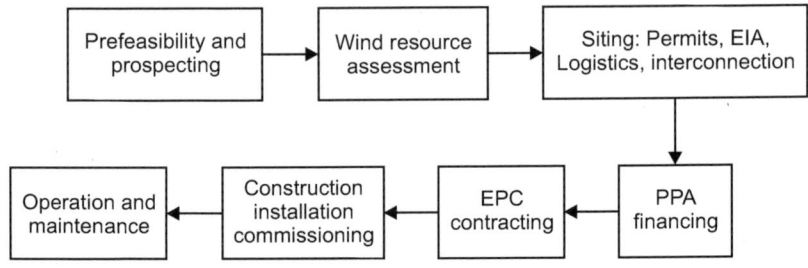

EIA = Environmental impact assessment
EPC = Engineering procurement and construction
PPA = Power purchase agreement

Figure 6.1: Activities in the life cycle of a wind energy project.

Table 6.1: Description of the activities in the life cycle of a wind energy project.

Activity (duration)	Description
Prefeasibility and prospecting (3 months)	The goal of prefeasibility analysis is to determine regions that have the potential for feasible wind projects. It uses publicly available wind resource maps along with considerations such as tariffs, land availability, licensing process, transmission, logistics, cost of project and others.
	Prospecting involves site visits to the regions that meet the prefeasibility requirement in order to collect additional data about terrain, vegetation, landownership, and other factors. The outcome of prospecting is a short-list of regions with the highest prospects and locations for wind measurement in these regions.
Wind Resource Assessment (WRA) (at least 15 months, 2–3 years for large wind power plants)	WRA is the quantification of wind resources to compute parameters such as average wind speed, average wind energy density and the average annual energy production (AEP) of a proposed wind power plant. A preliminary WRA is started in the previous activity. The bulk of the activities in this phase are wind measurement, AEP estimation based on wind flow modelling, AEP estimation based on computational fluid dynamics modelling for complex terrain and estimation of loss and uncertainty.
Siting: Permits, Environmental Impact Assessment (EIA), Logistics, Grid interconnection (6–12 months)	Project siting encompasses all the activities that precede construction, such as EIA, obstruction to aviation airspace analysis, interference with radar and telecommunications analysis, visual impact assessment, site planning and engineering, logistics planning, grid interconnection, and permitting and licensing. This activity may occur in parallel with the WRA step.

(Cont'd...)

Activity (duration)	Description
Power Purchase Agreement (PPA), Financing (3–6 months)	The focus of this activity is on the financial aspects of a wind project. PPA negotiation with a purchasing utility is the first step followed by agreements with equity and debt financiers. This activity occurs after the WRA and siting activities.
Engineering Procurement and Construction (EPC) Contracting (3 months, turbine delivery: 9 months)	This activity starts with the selection of an EPC contractor and selection of turbines for the project. In this activity, detailed project plans are created along with detailed engineering of the foundation, roads, collection system, substation, grid interconnection, logistics and others.
Construction, installation and commissioning (1 turbine/month)	In this activity, the wind power plant site is prepared, roads are constructed, foundations of turbines are constructed, towers are erected, the nacelle and rotor are lifted and the turbines are commissioned.
Operation and maintenance (ongoing)	After a wind farm is commissioned, ongoing operation and maintenance activities are performed.

6.4 Importance of accurate wind resource assessments

The importance of accuracy in wind resource may be illustrated through these two cases:

1. Inaccuracy of ±5% in wind measurement can lead to approximately ±10% inaccuracy in AEP, which in turn will lead to ±10% inaccuracy in revenue and finally this inaccuracy can be the difference between a financially viable and a financially unviable wind project. Hence, a lot of emphasis is placed on accurate wind speed measurements because this translates into higher accuracy in the estimate of a project's financial returns, a key requirement of financiers.

2. WPP projects are often delayed or rejected by financiers not because the wind resource is lacking, but because the WRA was conducted with insufficient rigor with respect to technical due diligence on measurement instruments, measurement setup, data management, wind flow modelling and assumptions.

This section will highlight the challenges associated with performing an accurate WRA and the impact on the financial feasibility of a wind project.

6.4.1 Wind resource assessment challenges

The following are prominent WRA challenges:

1. Since wind is a weather phenomenon, there is inherent uncertainty in forecasting the long-term wind resource and the AEP over the lifetime

of a WPP. Wind speed measured during a whole year provides a snapshot, however, year-to-year variation in wind speed is impacted by El Niño, La Niña and a variety of other long-term cyclical phenomena. Extrapolation of a year of measurement wind data to the lifetime of a WPP (more than 20 years into the future) requires statistical models and assumptions.

2. Wind measurement requires precise instruments, precise placement and orientation of instruments, no deterioration in performance of sensors for multiple years and an auditable process of managing vast amount of data. This requires sufficient budget and a dedicated and qualified team to operate and maintain the measurement campaign.

3. The location of measurement met-masts and the location of turbines are not the same. In a typical wind farm, one measurement mast is installed for every 10–15 wind turbines. Depending on the terrain, the difference in wind speeds between measured and turbine locations can be ±10%. In addition, the measurement height and the turbine hub height are typically not the same. Extrapolating wind speed from the measurement location to turbine locations and from measurement height to hub height requires wind flow modelling. Wind flow models are one of the most complex computational models in physics. Solving these models requires simplifying assumptions such as linearity, thermal stability, simple terrain, etc. All these factors introduce uncertainty in estimates of average wind speed at hub height and average AEP.

6.4.2 Impact of wind resource assessments on financial feasibility

The WRA is the core activity that determines the viability of a wind project. In simple terms, there are two impacts of the WRA on the financial feasibility of a project:

1. The primary output of a WRA is the AEP of a project, which directly influences the revenue of a project. If the AEP is low, then the revenue is low, hence, the project is less attractive.

2. The next most important output of a WRA is the uncertainty associated with AEP, which directly influences the risk of a project.

6.4.3 Impact of wind resource assessments on grid integration

Before a wind project is connected to the grid, a detailed system impact study is performed to understand the impact on the grid of connecting variable power from a wind project. A system impact study involves load flow, short-circuit and dynamic stability analyses.

The analyses require an accurate time series of 10-minute AEP data for at least 1 year, which is supplied by the WRA. The output of the analyses is a list of upgrades required in the grid to safely and reliably inject wind energy into the grid. Note for wind power penetration of less than 5% of total generation, typically there are no upgrades or minimal upgrades. In case upgrades are required, one or more of the following are on the list: transmission system, substations, protection systems and control systems of existing generators and new generators.

The second major impact of wind power is to system operations. System operations is responsible for minute-to-minute, hour-to-hour and day-to-day balancing of system wide load and energy production. To comprehend the impact of the WRA on system operations, consider the following situation. When load is decreasing, conventional power plants reduce production. Since the marginal cost of wind is close to zero, its merit order is the highest, which means all the wind energy should be dispatched.

However, there comes a point when the production at all the conventional power plants is at the lower limit and load is still falling. In such extreme situations, wind production may be curtailed. The amount of annual curtailment is important to financiers, utilities, developers and policy makers, it is predicted by running a simulation of the network with data about load, generators and 10 minute AEP time series data from the WRA.

6.4.4   Small and medium-sized wind projects

Nevertheless, the guidelines apply to small and medium-sized wind projects as well. For such projects, it is common to find cases in which one or more of the following are not performed. On-site measurement close to hub height using quality instruments, wind flow modelling using GIS model of terrain and roughness, long-term correction of AEP and uncertainty estimates of AEP. Such poorly conducted WRAs most certainly lead to inability to determine the financial viability of these projects. The result is a rude shock after project commissioning at the significantly lower energy than that predicted.

This is one of the primary causes of abandonment of a large fraction of small and medium sized wind projects—the actual revenue generated during operations is unable to cover the cost of operation and maintenance. It is, therefore, imperative that a rigorous WRA be performed for all wind projects regardless of size.

6.5   Wind resource assessment process

This section discusses the life cycle of WRA. The individual stages are briefly described in Table 6.2.

Table 6.2: Description of the five stages of a wind resource assessment.

Stages of a wind resource assessment	Description
Stage I: Prospecting	Prospecting begins with high-level analysis of wind-rich areas. Two useful sources of high-level wind data are 3 Tier and AWS true power. Both provide wind data based on numerical weather prediction mesoscale model and reanalysis data. These data are used to perform a preliminary wind resource assessment (WRA) with the goal of comparing areas based on annual energy production (AEP). At this stage, the AEP accuracy is ±50%.
Stage II: Wind measurement	At least 1 year of on-site wind measurement is required in the project area. Traditional met-masts and remote wind measurement are two methods.
Stage III: Linear wind flow model based estimation of AEP	Linear wind flow models such as WAPs are used for the core spatial extrapolation and AEP computation. Other software tools are Wind PRO, Wind Farmer and Open wind. The spatial and vertical extrapolations performed here are valid for mildly changing terrain, thermal stability (no convection) and terrain in which roughness around the measured sites are similar to roughness around the turbine sites.
Stage IV: CFD model based estimation of AEP	Computational fluid dynamics (CFD) modelling is performed if assumptions of the stage III model are not valid. It is required for cases with complex terrain and thermal instability. CFD is modelled on a three dimensional grid and these models are run on high performance computer clusters. Although CFD models add cost to a WRA, it is worth the investment when assumptions related to stage III of a WRA are not completely valid.
Stage V: Loss and uncertainty estimation	This is an important component of a WRA.

6.5.1 Loss and uncertainty estimation

Financiers of a wind project are keen to understand the sources and amount of both losses and uncertainty. Although losses and uncertainty are often mentioned together, these are mutually exclusive concepts.

Losses are estimates of a decrease in energy output that is known. As an example, 6% is the estimate of energy loss due to wake. This is one component of the estimated loss. Other sources of losses are electrical, plant availability, turbine performance, environmental and curtailment. In developed markets and in regions with well-known wind conditions, total losses are in the range of 10–12%. For complex projects, total losses can be in the vicinity of 15%. Losses are taken out of AEP estimates from stages III and IV to yield net AEP.

Uncertainty, on the other hand, is a statistical concept that describes the variability associated with an estimate. As an example, consider a case with an estimate of net AEP of 100 gigawatt-hours and uncertainty of 10% as shown in Fig. 6.2. Several factors may cause uncertainty in the estimate, which is measured in terms of standard deviation of net AEP. Common sources of uncertainty are wind speed measurement, wind speed extrapolation (spatial, vertical and temporal), power curve, wake, air density and others. In developed markets and regions with well-known wind conditions, total uncertainty (measured in terms of standard deviation of AEP as a percentage of average AEP) is about 12% and for newer markets, it may be 20%–25%. AEP for different exceedance probabilities are computed by subtracting appropriate multiples of standard deviation from net AEP, as is illustrated in Fig. 6.2.

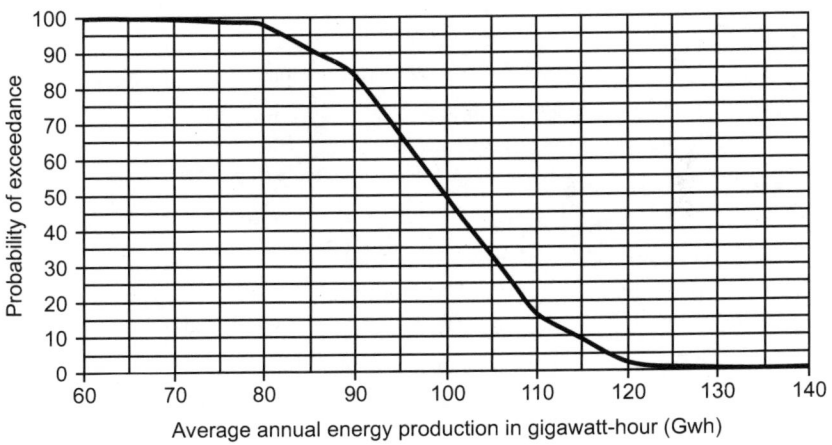

Figure 6.2: Illustration of exceedance probabilities.

6.5.2 Bankable wind resource assessments

The output of the WRA process is a bankable WRA. The concise description is: 'A bankable resource estimate is one in which enough (high quality) verifiable data is available to quantify the uncertainty in wind resource at the planned wind project location. A non-bankable WRA is normally the first stage of resource estimation when not all the uncertainties in wind conditions are completely understood or quantified and therefore, the return on investment has a high degree of uncertainty'.

6.6 Best practices for wind resource assessments

The focus of this section will be on WRA best practices for emerging wind energy markets. The best practices cover three areas: policy support for WRA, wind measurement and management of resource assessment uncertainty.

6.6.1 Policy

Feed-in Tariff

During the developing phase of the wind industry, the most important element of policy is tariffs. A feed-in tariff (FiT) is the most widely used and it is accepted as the most effective model for setting a tariff for renewable energy. Fit must be at a level that reflects both the wind resources present in the country and the anticipated cost of a wind project.

Licensing guidelines

After the precondition of a sufficiently high level of FiT is met, a policy prescription that can significantly reduce the lead time of wind projects is licensing guidelines that stipulate a minimum set of requirements that the WRA of a project should meet in order to qualify for a generation license:

1. At least 1 year of on-site wind measurement data should be collected with a met-tower of at least 60 meters height that is equipped with calibrated and high-quality anemometers at three heights.
2. The AEP should be based on a linear wind flow model for a simple terrain or a Computational Fluid Dynamics (CFD) model for a complex terrain.
3. Long-term correction of AEP should be computed to account for annual variability of the wind climate.
4. Site-specific losses and uncertainty should be estimated.
5. Extreme wind speed should be estimated to select a class of turbines suitable for the site.

The licensing guidelines should encapsulate the best practices of WRAs by requiring a detailed WRA report that contains the above mentioned components. Such requirements not only cull out weak developers but also attract foreign and domestic capital because financiers have a higher degree of confidence in the readiness of a project.

6.6.2 Grant or low-interest loan for wind resource assessments

The next policy prescription to accelerate wind development is to provide grants or low-interest loans to qualified developers for the WRA stage of a wind project. For a first-time wind project developer, the WRA activity may seem too long and expensive. The typical length of a WRA for a medium-sized wind farm of 20 megawatts should be approximately 2 years.

The grant/loan programme may be used as a vehicle to improve the quality of wind measurement by private developers. The following is a list of possible avenues:

1. Use international wind energy experts to review the grant/loan applications.
2. Publish a checklist that covers the following aspects of a measurement campaign: Location, personnel, instruments, installation and operation and maintenance. A grant/loan application should be required to follow the checklist.
3. Require a meteorologist or climate scientist on the team with experience in on-site measurements.
4. Encourage partnership with in-country research organisations, with the goals of both injecting scientific rigor in measurement and enhancing the capabilities of the scientific community within the country.

In order to put wind projects on the correct path during the nascent stage, a seed investment fund should be set up to invest in high-quality bankable WRAs. This fund should mandate that:

1. Project developers pursue a high-quality measurement campaign.
2. Project developers share wind measurement data for the purposes of enhancing country wide wind resource maps.

Sharing of wind data requires special mention in emerging wind energy markets. The lack of high-quality wind resource maps that are validated with high-quality measurements is a serious impediment to sustained growth of the wind industry. As a result of lack of high-quality wind resource maps, the cost of wind development remains high and the lead time from concept to commissioning of wind projects stays long.

However, for competitive reasons, project developers are unwilling to share wind measurement data, a precious commodity. There are several methods to approach the problem of sharing wind data. Similar approaches should be part of a grant/loan programme to increase the accuracy of wind resource maps through on-the-ground measurements.

6.6.3 Country wide resource assessment

A Wind Resource Map (WRM) for a country is the starting point for all wind development. Developers rely on WRM for prospecting and policy makers use it for developing wind energy road maps. Therefore, a high-quality WRM is crucial to lowering the cost and duration of WRA. Since wind is a localised phenomenon, an effective approach to planning and implementing wind energy policies is to define zones or corridors for wind development.

Wind energy corridors are regions screened for wind resources, land availability, infrastructure and other factors with the goal of focusing the efforts of all the agencies involved in licenses, approvals and permits to the identified regions. Policy makers and implementation agencies should fund wind corridor-specific: (i) high-quality, long-term meteorological measurements as described above, (ii) research and development of wind flow modelling and (iii) data collection and modelling of extreme wind speed events.

Such initiatives would lower the uncertainty of all projects in the corridor in three key areas: (i) long-term correction of AEP, (ii) extrapolation of wind speed to the entire project area and to hub height and (ii) choice of the most appropriate class of turbine as opposed to the most conservative class of turbine. This investment is not project specific, rather it significantly benefits the entire wind industry in the country.

6.6.4   Project-specific measurement

Project-specific on-site measurement is the most time-consuming and may be the most expensive activity in the WRA. When done with rigor, project-specific measurement may lead to a high-quality bankable WRA. The properties of a high-quality measurement campaign are given below:

1. Use a high-quality (preferably Class I anemometer) calibrated anemometer, as close to hub height as possible (preferably >2/3 of hub height). Install anemometers, preferably at three heights, so that vertical extrapolation may be accurately performed.

2. Use redundant anemometers so that potential for loss of data due to tower shadow or sensor failure is minimal. Use long booms to minimise the impact of flow distortion.

3. Deploy two or more met-masts for a wind-farm site, preferably one met-mast for every 5–8 turbines or 10–20 MW capacity (the lower number is for a complex terrain and the higher number for a simple terrain).

4. Collect and analyse daily data feeds rigorously. Ensure that raw data are archived and an audit trail exists for data corrections, so that the data can be independently verified.

5. Collect data for at least 1 year, if measurement is done for more than 1 year, then collect data for a full 2 or 3 years.

6.6.5   Managing resource assessment uncertainty

Often a WRA is non-bankable, not because the wind resource is lacking, but because the uncertainty associated with AEP is high. High uncertainty in resource assessment is due to insufficient rigor with respect to technical due diligence on measurement, data, wind flow models and assumptions.

The following points highlight the description of the detailed impact of higher uncertainty, related to WPP project:

1. Higher uncertainty leads to lower power which leads to lower revenue. And thus, its leads to lower cash flow, which in turn leads to lower return on equity.

2. Lower cash flow leaves a smaller cushion for payment of principal and interest on debt. In other words, the Debt Service Coverage Ratio (DSCR) is lower. Most debt issuers expect the minimum DSCR for a project to be in the range of 1.25 to 1.3. When this minimum DSCR cannot be met, a project is forced to reduce the amount of debt, which means higher equity is required, which leads to lower return on equity.

The double impact on return on equity as described often causes a project to be one of the following:

1. Delayed while the project developer attempts to reduce uncertainty. Reduction in uncertainty may be accomplished through additional years of measurement with higher-accuracy sensors or installation of new met-masts. This results in higher investment and longer lead time for projects.

2. Abandoned because the developer does not wish to invest more to continue the development phase of the project.

In order to avoid such scenarios, uncertainty should be managed from the very first day. The measurement practices mentioned in the previous section contribute greatly to lowering of uncertainty. Other large sources of uncertainty arise from terrain complexity. For complex terrains, this uncertainty may be reduced by increasing the number of measurement stations, deploying remote sensing instruments and using CFD wind flow models for extrapolation.

There are five primary stakeholders of the WRA: policy makers, private developers, implementation agencies, donors and financing agencies and utilities. The best practices presented above are summarised in Table 6.3 from the perspective of the five classes of stakeholders.

Table 6.3: Best practices from the perspectives of the five stakeholders.

Stakeholders	Best practices
Policy makers	Licensing guidelines: Detailed licensing guidelines should be developed with requirements about duration and quality of measurement and quality of wind flow modelling and data processing.
	Country-wide resource assessment: Fund research and development on improving predictability of wind flow models
	Long-term (10–20 years) measurement for high-quality reference wind data: Fund ongoing research and development of long-term measurement campaigns in resource rich areas.

(Cont'd...)

Stakeholders	Best practices
	Grant or low-interest loan for wind resource assessments: Develop fund (grants, low-interest loans) to fund development of bankable wind resource assessments.
Private developers	Managing resource assessment uncertainty: Higher quality measurement and wind flow modelling.
	Project-specific measurement: Implement international best practices and perform with rigor.
Implementation agencies	Licensing guidelines: Coordinate among all agencies involved with licensing, permitting and approving wind projects to develop transparent and rigorous criteria.
	Country wide resource assessment: Fund research and development on improving predictability of wind flow models.
Donors/ financing agencies	Grant or low-interest loan for wind resource
	Assessments: Fund development of high quality bankable wind resource assessments.
	Long-term (10–20 years) measurement for high-quality: Fund long-term measurement campaigns.
Utilities	Country wide resource assessment: Basis for corridorwide system impact study and planning for construction or upgrade of transmission lines and associated infrastructure.
	Project-specific measurement: Accurate estimate of annual energy production, which forms the basis for wind project-specific grid impact study.

To sum up, WRA is a pivotal step in the development phase of a wind project. Even if the wind resource is sufficient for a viable project, the difference between a bankable and non-bankable WRA is the amount of due diligence exercised during the performance of wind measurement and wind flow modelling to compute AEP. With respect to due diligence, financiers look for the developer's efforts to reduce uncertainty during the WRA and the developer's ability to estimate the remaining uncertainty.

Recent findings suggest that in countries with emerging wind energy markets, a large fraction of developers: (i) do not conduct WRAs with sufficient due diligence and (ii) do not adequately understand the concept of uncertainty and its implications on the bankability of a WRA.

There are various reasons: Ignorance about requirements of bankable WRAs and/or lack of funds to invest in the development activity. Key policy support required for WRAs, to enable accelerated wind development, are the following:

1. Rigorous licensing preconditions for wind farms, with clearly stated requirements and checklist for WRAs, with an intent to pro-actively inform developers about the need for quality and rigor in bankable WRAs.

2. Low-cost financing or grants for qualified wind energy developers to perform WRAs.
3. Long-term quality wind measurement campaign in order to:
   (a) Collect reference data, which will reduce uncertainty.
   (b) Map the wind resources to a greater level of detail in areas with promising wind resources.
4. Capacity building to transfer knowledge about best practices of WRAs.

Offshore wind energy

7.1   Introduction

The potential for offshore wind is enormous in Europe and elsewhere, but the technical challenges are also great. The capital costs are higher than onshore, the risks are greater, the project sizes are greater and the costs of mistakes are greater. Offshore wind technology and practice has come a long way in a short time, but there is clearly much development still to be done.

Although the fundamentals of the technology are the same onshore and offshore, it is clear that offshore wind technology is likely to diverge further from onshore technology. Methods of installation and operation are already very different from onshore wind generation, with great attention being given to reliability and access.

7.2   Wind resource assessment offshore

This section describes the differences in wind flow, monitoring and data analysis offshore in comparison to onshore. It also highlights the key differences associated with the assessment of the offshore wind resource and the energy production of offshore wind farms when compared with onshore wind farms. Many of the elements of the analyses are common to onshore and offshore projects and it is therefore recommended that this section is read in parallel with the chapters on onshore wind.

Onshore, topographic effects are one of the main driving forces of the wind regime. With no topographic effects offshore, other factors dominate the variation in wind speed with height. The surface roughness (a parameter used to describe the roughness of the surface of the ground, referred to as Z_0) is low, which results in a steeper boundary layer profile (also referred to as the wind shear profile), characterised by the symbol α. A range of typical values for Z_0 and α are illustrated in Table 7.1. Offshore, the surface roughness length is dependent on sea state, increasing with the local wave conditions, which are in turn influenced by wind conditions. However, this relationship is complex, as the sea surface, even when rough, does not present fixed roughness elements such as trees, hills and buildings, as tends to be the case onshore. The low surface roughness also results in low turbulence intensity. This serves to reduce mechanical loads. It also may increase the energy capture compared to an identical wind turbine at an onshore location with identical mean wind speed.

The coastal zone, where the properties of the boundary layer are changeable, extends away from the shore for varying distances and this can result in variations in wind speed, boundary layer profiles and turbulence across the wind farm.

Table 7.1: Typical values for Z_0 and α.

Type of terrain	Z_0 (m)	α
Mud flats, ice	0.00001	
Smooth sea	0.0001	
Sand	0.0003	0.10
Snow surface	0.001	
Bare soil	0.005	0.13
Low grass, steppe	0.01	
Fallow field	0.03	
Open farmland	0.05	0.19
Shelter belts	0.3	
Forest and woodland	0.5	
Suburb	0.8	
City	1	0.32

Stable flow conditions are also evident offshore. In these situations, air flows with different origins and air temperatures can be slow to mix. This can manifest itself as unusual boundary layer profiles and in some rare situations wind speed may even reduce with height.

A further factor influencing offshore winds can be the tide level in areas with a high tidal range. The rise and fall of the sea level effectively shifts the location of the turbine in the boundary layer. This can have impacts in variation of mean wind speed within a period of approximately 12 hours and also on the variation in mean winds across the turbine rotor itself. Temperature-driven flows due to the thermal inertia of the sea initiate localised winds around the coastal area. Compared to the land, the sea temperature is more constant over the day. During the day, as the land heats up, the warmer air rises and is replaced by cooler air from over the sea. This creates an onshore wind. The reverse effect can happen during the night, resulting in an offshore wind. The strength and direction of the resulting wind is influenced by the existing high-level gradient wind and in some situations the gradient wind can be cancelled out by the sea breeze, leaving an area with no wind. Finally, as all sailors are aware, close to the coast there are 'backing' and 'veering' effects.

7.2.1 Measurement of offshore wind farm

Offshore wind farms typically use the largest available wind turbines on the market. Their size presents several issues, including understanding the characteristics of the boundary layer up to and above heights of 100 m. Measurements offshore are costly, with costs driven to a large extent by the cost of constructing the support structure for the meteorological mast. Offshore monitoring towers are un-guyed and therefore need to be wider, which can mean measurements are more susceptible to wind flow effects from the tower. Anemometry equipment is otherwise standard.

If high-quality wind measurements are not available from the site or nearby, there are other sources of information that can be utilised to determine the approximate long-term wind regime at the wind farm location. There are offshore databases for wind data, including meteorological buoys, light vessels and observation platforms. Additionally, meso-scale modelling (based on global reanalysis data sets) and Earth-observation data play a role in preliminary analysis and analysis of spatial variability.

7.2.2   Wind analysis offshore

Depending on the amount of data available, different analysis methods can be employed. A feasibility study can be carried out based on available wind data in the area. WAsP can be used from coastal meteorological stations to give a prediction offshore, which is aided by its latest tool, the Coastal Discontinuity Model (CDM).

Existing offshore measurements can also be used. There are problems associated with using long-distance modelling, however, especially around the coast, due to the differences in predominant driving forces between onshore and offshore breezes and the variation in the coastal zone in between.

For a more detailed analysis, measurements offshore at the site are necessary. Measure Correlate Predict (MCP) methods from a mast offshore to an onshore reference station can be used. With several measurement heights and attention to measurement, more accurate modelling of the boundary layer will help extrapolate to heights above the monitoring mast.

7.3   Energy prediction for offshore wind turbines

The energy prediction step is essentially the same as for onshore predictions. There is generally only minor predicted variation in wind speed over a site. Given the absence of topography offshore, measurements from a mast can be considered representative of a much larger area than would be possible onshore.

For large offshore sites, wake losses are likely to be higher than for many onshore wind farms. The wake losses are increased due to the size of the project and also due to lower ambient turbulence levels – the wind offshore is much smoother. There is therefore less mixing of the air behind the turbine, which results in a slower re-energising of the slow moving air and the wake lasts longer. Observations from the largest current offshore wind farms have identified shortcomings in the classic wind farm wake modelling techniques, due to the large size of the projects and perhaps due to specific aspects of the wind regime offshore. Relatively simple amendments to standard wake models are currently being used to model offshore wake effects for large projects, but further research work is ongoing to better understand the mechanisms involved and to develop second generation offshore wake models.

There is likely to be more downtime of machines offshore, primarily due to difficult access to the turbines. If a turbine has shut down and needs maintenance work, access to it may be delayed until there is a suitable window in the weather conditions. This aspect of offshore wind energy is a critical factor in the economic appraisal of a project. Increasingly sophisticated Monte Carlo-based simulation models are being used to assess the availability of offshore wind farms, which include as variables the resourcing of servicing crews, travel time from shore, the turbine technology itself and sea state.

7.4　Wind turbine technology for offshore locations

7.4.1　Availability, reliability and access

High availability is crucial for the economics of any wind farm. This depends primarily on high system reliability and adequate maintenance capability, with both being achieved within economic constraints on capital and operational costs. Key issues to be addressed for good economics of an offshore wind farm are:

1. Minimisation of maintenance requirements.
2. Maximisation of access feasibility.

The dilemma for the designer is how best to trade the cost of minimising maintenance by increasing reliability – often at added cost in redundant systems or greater design margins – against the cost of systems for facilitating and increasing maintenance capability. Access is critical as, in spite of the direct cost of component or system replacement in the difficult offshore conditions, lost production is often the greatest cost penalty of a wind turbine fault. For that reason much attention is given to access. Related to the means of access is the feasibility of various types of maintenance activities and whether or not support systems (cranes and so on) and other provisions are needed in the wind turbine nacelle systems.

Impact on nacelle design

The impacts of maintenance strategy on nacelle design relate to:

1. Provision for access to the nacelle.
2. Systems in the nacelle for handling components.

3. The strategic choice between whether the nacelle systems should be: (a) designed for long life and reliability in an integrated design that is not particularly sympathetic to local maintenance and partial removal of subsystems or (b) designed in a less cost-effective modular way for easy access to components.

Location of equipment

Transformers may be located in the nacelle or inside the tower base. Transformer failures have occurred in offshore turbines, but it is not clear that there is any fundamental problem with location in either the nacelle or the tower base.

Importance of tower top mass

The tower top mass is an important influence on foundation design. In order to achieve an acceptable natural frequency, greater tower top mass may require higher foundation stiffness, which could significantly affect the foundation cost for larger machines.

Internal cranes

One option is to have a heavy duty internal crane. Some companies have adopted an alternative concept, which in general consists of a lighter internal winch that can raise a heavy duty crane brought in by a maintenance vessel. The heavy duty crane may then be hoisted by the winch and set on crane rails provided in the nacelle. Thus it may be used to lower major components to a low-level platform for removal by the maintenance vessel.

Critical and difficult decisions remain about which components should be maintained offshore in the nacelle, which can be accessed, handled and removed to shore for refurbishment or replacement and when to draw a line on component maintenance capability and accept that certain levels of fault will require replacement of a whole nacelle.

Means of access

The costs of turbine downtime are such that an effective access system offshore can be relatively expensive and still be justified. Helicopter access to the nacelle top has been provided in some cases. The helicopter cannot land, but can lower personnel. Although having a helipad that would allow a helicopter to land is a significantly different issue, the ability to land personnel only on the nacelle top of a wind turbine has very little impact on nacelle design.

Access impediments: In the Baltic Sea especially, extensive icing occasionally takes place in some winters. This changes the issues regarding access, which may be over the ice if it is frozen solid or may use ice-breaking ships. Denmark and Sweden, provides for a section at water level with a bulbous shape. This assists in ice breaking and easing the flow of ice around the wind turbine, thereby reducing loads that would tend to move the whole foundation. In the European sites of the North Sea, the support structure design conditions are more likely to relate to waves than ice and early experience of offshore wind has shown clearly that access to a wind turbine base by boat is challenging in waves of around 1 m height or more.

Access technology development

There may be much benefit to be gained from the general knowledge of offshore industries that are already developed, especially the oil and gas industry. However, there are major differences between an offshore wind farm and for example, a large oil rig. The principal issues are:

1. There are multiple smaller installations in a wind farm and no permanent (shift-based) manning, nor the infrastructure that would necessarily justify helicopter use.

2. Cost of energy rules wind technology, whereas maintenance of production is much more important than access costs for oil and gas.

Thus, although the basis of solutions exists in established technology, it is not the case that the existing offshore industry already possesses off-the-shelf solutions for wind farm construction and maintenance. This is evident in the attention being given to improved systems for access, including the development of special craft.

In summary, there is an appreciable benefit in increasing accessibility rates upwards from the current 1.2 to 1.5 m threshold to 2.0 m. This has proven to be achievable using catamarans. There are a number of alternative vessels at the concept stage that are being designed to allow safe transfer in 2 m significant waves. It may be possible to achieve further improvements in accessibility using specialised workboats fitted with flexible gangways that can absorb the wave energy that cannot be handled by the vessel. Alternatively, larger vessels with a greater draught, which are more inherently stable in rougher sea conditions, must be employed. However, the drawback for offshore wind energy is that foundation technology and the economics for many projects dictate that turbines should be installed in shallow banks, hence shallow draught service vessels are obligatory.

7.4.2 Lightning risk offshore

Lightning has been more problematic offshore than expected. The answer, however, lies with providing wind turbine blades with better methods of lightning protection, as used on the more problematic land based sites, rather than looking for systems to ease handling and replacement of damaged blades.

The consequences of lightning strikes can be severe if systems are not adequately protected.

7.4.3 Maintenance strategy–Reliability versus maintenance provision

Regarding cost of energy from offshore wind, a general view is emerging that it is a better to invest in reliability to avoid maintenance than in equipment to facilitate it. Also, expenditure on maintenance ships and mobile gear is generally more effective than expenditure per machine on added local capability such as nacelle cranes. In selected cases, a strategy to facilitate *in situ* replacement of life-limited components (such as seals) may be advised.

7.5 Wind farm design offshore

Designing an offshore wind farm is a staged process involving:
1. Data-gathering.
2. Preliminary design and feasibility study.
3. Site investigation.
4. Concept development and selection.
5. Value engineering.
6. Specification.
7. Detailed design.

Close interaction throughout the design process is required, including interaction about the grid connection arrangements and constraints introduced through the consenting process. Key aspects of the design and the process of arriving at that design, are described in the following sections. The capital cost of offshore projects differs markedly from those onshore, with perhaps 50% of the capital cost being due to non-turbine elements, compared to less than 25% in onshore projects.

7.5.1 Site selection

The selection of the site is the most important decision in the development of an offshore wind farm. It is best accomplished through a short-listing process that draws together all known information on the site options, with selection decisions driven by feasibility, economics and programme, taking account of information on consenting issues, grid connection and other technical issues.

7.5.2 Wind turbine selection

Early selection of the wind turbine model for the project is typically necessary so that the design process for support structures (including site investigations), electrical system and grid connection can progress. Offshore projects require use of the larger wind turbines on the market, meaning that there is often limited choice, hence, securing the wind turbine model may be necessary to start up the project programme.

7.5.3 Layout

The process for designing the layout of an offshore wind farm is similar to the process for an onshore wind farm, albeit with different drivers. Once the site is secured by a developer, the constraints and known data on the site are evaluated and input in the layout design, as shown in Fig. 7.1.

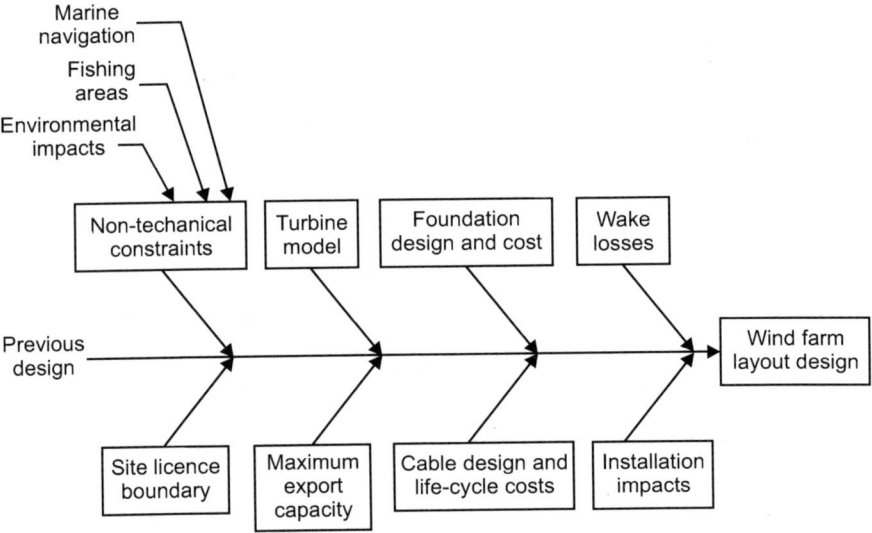

Figure 7.1: Layout design process.

One driver that often dominates onshore wind farm design is the noise footprint of the project, which is not usually an issue for offshore projects.

The layout design process evaluates and compares layout options in relation to technical feasibility, overall capital cost and the predicted energy production. Determining the optimum layout for an offshore wind farm involves many trade-offs. One example of such a trade-off is related to the array spacing, where a balance must be struck between array losses, that is to say, energy production and electrical system costs and efficiency. Several other such trade-offs exist.

The experience to date is that for sites of homogeneous depth and soil properties, revenue and hence production, has a dominant impact on the cost of energy and as a result the design of the layout is dictated by energy production. However, where the water depth and soil properties vary widely across a site, more complex trade-offs must be made between production, electrical system costs and support structure costs, including installation costs.

7.5.4 Offshore support structures

Support structures for offshore wind turbines are highly dynamic, having to cope with combined wind and hydrodynamic loading and complex dynamic behaviour from the wind turbine. It is vital to capture the integrated effect of the wind and wave loads and the wind turbine control system, as this is a situation where the total loading is likely to be significantly less than the sum of the constituent loads.

This is because the loads are not coincident and because the aerodynamic damping provided by the rotor significantly damps the motions due to wave loading. Structures to support wind turbines come in various shapes and sizes. Concrete gravity base structures have also been used on several projects. As wind turbines get larger and are located in deeper water, tripod or jacket structures may become more attractive. The design process for offshore wind turbine support structures is illustrated in Fig. 7.2.

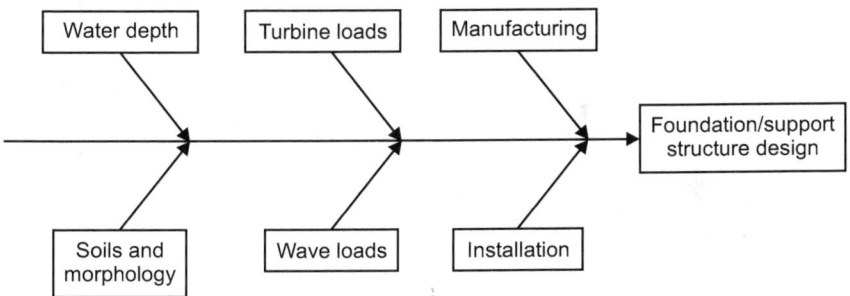

Figure 7.2: Typical primary support structure design inputs.

The structure designs are strongly influenced by metocean site conditions and site investigations. Metocean conditions are determined by detailed hydrodynamic analysis based on long-term hindcast model data and calibrated against short-term site wave measurements.

Site investigations are major tasks in their own right, requiring careful planning to achieve optimum results within programme and financial constraints. These involve a combination of geophysical and geotechnical measurements. The geotechnical investigations identify the physical properties of the soils into which foundations are to be placed and are achieved using cone penetrometer or borehole testing. Geophysical tests involve measurement of water depth and of the seismic properties of the underlying soil layers and can be used to interpolate the physical findings of geotechnical tests. The type and extent of geotechnical tests is dependent on the soil type occurring at the site and the homogeneity of those site conditions. Design of the secondary structures, such as decks, boat landings and cable J-tubes, is typically developed in the detailed design phase. These details have a major impact on ease of construction, support structure maintenance requirements, accessibility of the wind turbines and safety of personnel during the operations phase. Hence a significant design period is recommended.

Where an offshore substation is required, this is likely to require a substantial support structure, although as it does not have the complexities of wind turbine loading, it is a more conventional offshore structure to design. As discussed in the following section, an offshore substation may range from a unit of below 100 MW with a small single deck structure to a large, multi-tier high voltage DC (HVDC) platform.

7.5.5   Electrical system

An offshore wind farm electrical system consists of six key elements:

1. Wind turbine generators.
2. Offshore inter-turbine cables (electrical collection system).
3. Offshore substation (if present).
4. Transmission cables to shore.
5. Onshore substation (and onshore cables).
6. Connection to the grid.

Figure 7.3 shows the electrical network of offshore wind turbines.

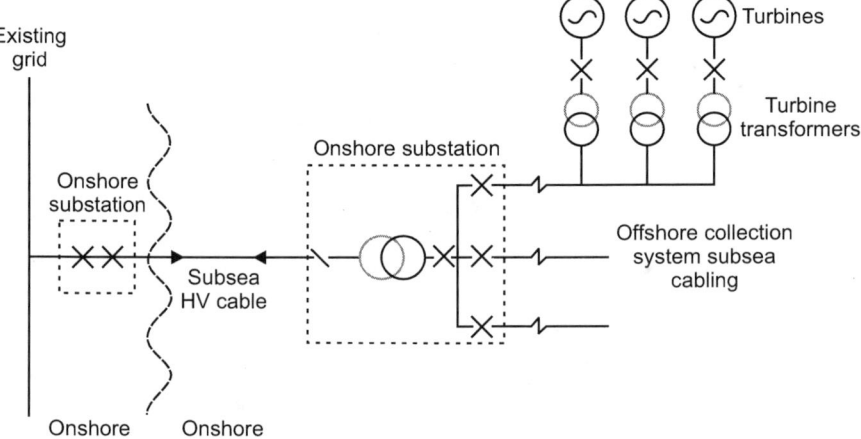

Figure 7.3: Electrical network of offshore wind turbines.

The design of the electrical system is determined by the characteristics of the wind turbine generators and of the network to which the project is to be connected, as well as regulations imposed upon it, notably through grid codes. The network operator controls the grid to meet its operational objectives and also requires a degree of control over large generators (which may include offshore wind farms). Additionally, the wind farm must be designed to respond appropriately to grid faults. These demands can be expected for any large wind farm located offshore. Wind turbine control and electrical systems are constantly evolving to provide improved characteristics and fault response for the purpose of grid integration. Nevertheless, the wind farm electrical system can be expected to have additional functional requirements in addition to the basic transmission from turbines to the grid connection point.

Offshore substations

Offshore substations are used to reduce electrical losses by increasing the voltage and then exporting the power to shore. Generally a substation does not need to be installed if:

1. The project is small (~100 MW or less).
2. It is close to shore (~15 km or less).
3. The connection to the grid is at collection voltage (for example 33 kV).

Most early offshore wind projects met some or all of these criteria, so were built without an offshore substation. However, most future offshore wind farms will be large and/or located far from shore and so will require one or more offshore substations.

Offshore substations typically serve to step up the voltage from the site distribution voltage (30–36 kV) to a higher voltage (say 100–220 kV), which will usually be the connection voltage. This step-up dramatically reduces the number of export circuits (subsea cables) between the offshore substation and the shore. Typically, each export circuit may be rated in the range 150–200 MW.

Onshore substations

Design of the onshore substation may be driven by the network operator, but there will be some choices to be made by the project developer. Generally, the onshore substation will consist of switchgear, metering, transformers and associated plant. The onshore substation may also have reactive compensation equipment, depending on the network operator requirements and the design of the offshore network.

Subsea cables

Subsea cables are of well-established design. Each circuit runs in a single cable containing all three phases and optical fibre for communications, with a series of fillers and protective layers and longitudinal water blocking to prevent extensive flooding in the event of the external layers failing.

Long-term reliability of the subsea cables is a major concern, addressed mainly by ensuring the safe burial of the cables at a depth that avoids damage from trawlers and anchors and the exposure of cables to hydrodynamic loading.

7.5.6   Installation of wind turbines

The installation of the wind turbines and their support structures is a major factor in the design of offshore wind farms, with the specific challenge of having to perform multiple repeated operations in difficult offshore locations. As well as being a significant contributor to the capital cost, the installation process may drive the selection of support structure technology.

The substation installation is also a major operation, albeit a single one, unlike the wind turbine and wind turbine foundation installation and therefore much more conventional in the offshore industry. A large crane vessel is likely to be required – either a shear-leg crane or other heavy lift unit. Alternatively, self-installing substation designs may find an increasing role in the future.

Cable installation is a significant industry sector, with specialist design and planning and installation vessels and equipment. The design and planning of the cable installation is an early activity, covering:

1. Identification of hazards to cables.
2. Site investigation to identify seabed properties (geophysical survey, vibrocore sampling, cone penetrometer tests, boreholes).
3. Development of burial protection indices.
4. Scour protection.
5. Cable route selection.
6. Cable transport.
7. Vessel and equipment selection.

The experience has shown that the sector presents unique technical challenges that must be addressed through research and development efforts:

1. Any project involves multiple distributed installations, spread over much larger areas and in much larger numbers than other offshore industries.
2. Nearshore shallow water (for most projects) siting – unlike oil and gas rigs, sea defence works and ports and harbours.
3. More stringent economics than oil and gas.

These factors combine so that there is limited borrowing available from other sectors and technology has had to evolve within a short timescale on a small number of projects, leaving significant scope for further maturing.

This issue covers all parts of the industry, including:

1. Wind turbines.
2. Wind turbine support structures.
3. Modelling tools.
4. Electrical infrastructure.
5. Assembly and installation.
6. Operations and maintenance.

Two drivers cut across all these areas: Safety of personnel and the public and environmental protection.

Some of the design drivers for a wind turbine installed offshore are fundamentally different from those installed onshore, specifically:

1. The non-turbine elements of an offshore project represent a much higher proportion of the capital cost, with that cost element only partially scaling with turbine size.
2. Acceptable noise levels are much higher offshore.
3. Better reliability is required offshore.

These drivers have already influenced the design of wind turbines used offshore and this is leading to the development of wind turbines specifically designed for offshore use with features such as:

1. Larger rotors and rated power.
2. Higher rotor tip speeds.
3. Sophisticated control strategies.
4. Electrical equipment designed to improve grid connection capability.

Wind resource assessment offshore provides more background, but other technological innovations that may be deployed in future offshore turbines include:

1. Two-bladed rotors.
2. Downwind rotors.
3. More closely integrated drive trains.
4. Multi-pole permanent magnet generators.
5. High temperature superconductors (in generators).
6. High voltage output converters (eliminating the need for turbine transformers).

7.5.7 Wind turbine support structures

The support structures form a significant proportion of offshore wind development costs. It is expected that there will be both innovation and value engineering of structure designs and improved manufacturing processes to improve the economics and meet the demands for more challenging future sites and wind turbines.

Such developments are likely to include modifications to conventional designs, scale-up of manufacturing capacity and processes and more novel design concepts. Such innovative designs may include:

1. Suction caisson monotowers.
2. Use of suction caissons as the foundation of jacket or tripod structures.
3. Application of screw piles.
4. Floating structures.
5. Braced supports to monopiles.

There are two key aspects to the maturing of offshore support structures:

1. Acquisition of data on the behaviour of the existing structures in order to support research into the development of improved design tools and techniques and better design standards, this will be used to extend the life of structures, to reduce costs and to develop risk-based life-cycle approaches for future designs.
2. The build-up of scale and speed in production in order to achieve cost reduction and the capacity necessary to supply a growing market

The development of floating structures, while long-term, will be a major advance if successful.

7.5.8   Modelling tools in offshore wind turbines

The offshore wind sector will deploy turbines of greater size and in greater numbers than has been done previously. Understanding of the engineering impacts of this is achieved through modelling and this increase in scale requires development and validation of the industry's modelling tools. Associated with this is the refinement of design standards.

The priority areas that must be addressed for large offshore wind farms are:

1. Development of wind turbine wakes within the wind farm.
2. Meso-scale modifications to the ambient flow in the immediate environs of the wind farm.
3. Downstream persistence of the modification to ambient flow and therefore the impact of neighbouring wind farms upon each other.
4. Dynamic loads on wind turbines deep within wind farms.

7.5.9   Electrical infrastructure

Incremental development in electrical equipment (switchgear, transformers and reactive power compensation equipment) is to be expected, driven by the wider electricity supply industry. The offshore wind business will soon be the largest market for subsea cables and so some innovations there may be driven by the specific requirements of the sector, although cables are a relatively mature technology. Voltage-source high voltage DC transmission is a relatively new commercial technology and one that will find extensive application in offshore wind. The major electrical impact of the offshore sector will be the reshaping of the transmission network of the countries involved in order to serve these major new generating plants. Also to be expected is an increase in the inter-connection of countries to improve the firmness of national power systems, which may also involve providing an international offshore transmission network dedicated to serving offshore wind projects.

7.5.10   Assembly and installation

Future technical developments in the construction process are likely to be:

1. Improvements in harbour facilities that are strategically located for the main development regions.
2. Construction of further purpose-built installation equipment for the installation of wind turbines, support structures and subsea cables: vessels and also piling hammers, drilling spreads and cable ploughs.
3. Development of safe, efficient, reliable and repeatable processes to reduce costs, minimise risks, guarantee standards and deliver investor confidence.

7.5.11   Operations and maintenance

Successful performance of O&M is most critically dependent on service teams being able to access the wind farm as and when needed. Good progress has been made on this in recent years and accessibility has improved significantly.

This has been achieved by incremental improvements in:

1. Vessels used.
2. Landing stages on the wind turbine structures.
3. Procedures.

Future offshore wind farms offer new access challenges, being larger and much further offshore. This will result in increased use of helicopters for transferring service crews, larger vessels to give fast comfortable transit from port to site and the use of offshore accommodation platforms, combined with evolution of strategies to perform O&M.

7.5.12   Floating systems

The main drivers for floating technology are:

1. Access to useful resource areas that are in deep water yet often near the shore.

2. Potential for standard equipment that is relatively independent of water depth and seabed conditions.
3. Easier installation and decommissioning.
4. The possibility of system retrieval as a maintenance option.

Thus, the main obstacle to the realisation of such technology is the development of effective design concepts and demonstration of cost-effective technology, especially in respect of the floater and its mooring system.

Wind farm

8.1   Introduction

As in most other areas of power production, when it comes to capturing energy from the wind, efficiency comes in large numbers. Groups of large turbines, called wind farms or wind plants, are the most cost-efficient use of wind energy capacity. The most common utility-scale wind turbines have power capacities between 700 KW and 1.8 MW and they're grouped together to get the most electricity out of the wind resources available. They are typically spaced far apart in rural areas with high wind speeds and the small footprint of Horizontal Axis Wind Turbines (HAWTs) means that agricultural use of the land in nearly unaffected. Wind farms have capacities ranging anywhere from a few MW to hundreds of MW. The world's largest wind plant is the Raheenleagh Wind Farm located off the coast of Ireland. At full capacity (it's currently operating at partial capacity), it will have 200 turbines, a total power rating of 520 MW and cost nearly $600 million to build.

8.2   Factors affecting turbine location

Once a site has been identified and the decision has been taken to invest in its development, the wind farm design process begins. The fundamental aim is to maximise energy production, minimise capital cost and operating costs and stay within the constraints imposed by the site. As the constraints and costs are all subject to some level of uncertainty, the optimisation process also seeks to minimise risk. The first task is to define the constraints on the development:

1. Maximum installed capacity (due to grid connection or power purchase agreement terms).
2. Site boundary.
3. 'Set back'–distances from roads, dwellings, overhead lines, ownership boundaries and so on.
4. Environmental constraints.
5. Location of noise-sensitive dwellings, if any and assessment criteria.
6. Location of visually-sensitive viewpoints, if any and assessment criteria.
7. Location of dwellings that may be affected by 'shadow flicker' (flickering shadows cast by rotating blades) when the sun is in particular directions and assessment criteria.

8. Turbine minimum spacings, as defined by the turbine supplier (these are affected by turbulence, in particular).

9. Constraints associated with communications signals, for example microwave link corridors or radar.

These constraints may change as discussions and negotiations with various parties progress, so this is inevitably an iterative process.

When the likely constraints are known, a preliminary design of the wind farm can be produced. This will allow the size of the development to be established. As a rough guide, the installed capacity of the wind farm is likely to be of the order of 12 MWperkm, unless there are major restrictions that affect the efficient use of the available land.

For the purpose of defining the preliminary layout, it is necessary to define approximately what sizes of turbine are under consideration for the development, as the installed capacity that can be achieved with different sizes of turbine may vary significantly. The selection of a specific turbine model is often best left to the more detailed design phase, when the commercial terms of potential turbine suppliers are known. Therefore at this stage it is either necessary to use a 'generic' turbine design, defined in terms of a range of rotor diameters and a range of hub heights, or alternatively to proceed on the basis of two or three layouts, each based on specific wind turbines.

The preliminary layout may show that the available wind speed measurements on the site do not adequately cover all the intended turbine locations. In this case it will be necessary to consider installing additional anemometry equipment. The preliminary layout can then be used for discussions with the relevant authorities and affected parties. This is an iterative process and it is common for the layout to be altered at this stage.

The factors most likely to affect turbine location are:

1. Optimisation of energy production.
2. Visual influence.
3. Noise.
4. Turbine loads.

8.3 Optimisation of energy production

Once the wind farm constraints are defined, the layout of the wind farm can be optimised. This process is also called wind farm 'micro-siting'. As noted above, the aim of such a process is to maximise the energy production of the wind farm whilst minimising the infrastructure and operating costs. For most projects, the economics are substantially more sensitive to changes in energy production than infrastructure costs. It is therefore appropriate to use energy production as the dominant layout design parameter.

The detailed design of the wind farm is facilitated by the use of Wind Farm Design Tools (WFDTs). There are several that are commercially available and others that are research tools. Once an appropriate analysis of the wind regime at the site has been undertaken, a model is set up that can be used to design the layout, predict the energy production of the wind farm and address issues such as visual influence and noise.

For large wind farms it is often difficult to manually derive the most productive layout. For such sites a computational optimisation using a WFDT may result in substantial gains in predicted energy production. Even a 1% gain in energy production from improved micro-siting is worthwhile, as it may be achieved at no increase in capital cost. The computational optimisation process will usually involve many thousands of iterations and can include noise and visual constraints. WFDTs conveniently allow many permutations of wind farm size, turbine type, hub height and layout to be considered quickly and efficiently, increasing the likelihood that an optimal project will result. Financial models may be linked to the tool so that returns from different options can be directly calculated, further streamlining the development decision-making process.

8.3.1   Visual influence of wind turbines

'Visual influence' is the term used for the visibility of the wind turbines from the surrounding area. In many countries the visual influence of a wind farm on the landscape is an important issue, especially in regions with high population density. The use of computational design tools allows the Zone of Visual Influence (ZVI), or visibility footprint, to be calculated to identify from where the wind farm will be visible. It is usually necessary to agree a number of cases with the permitting authorities or other interested parties, such as:

1. Locations from which 50% of turbine hubs can be seen.
2. Locations from which at least one hub can be seen.
3. Locations from which at least one blade tip can be seen.

It is also common to generate 'visualisations' of the appearance of the wind farm from defined viewpoints. These can take the form of 'wireframe' representations of the topography. With more work, photomontages can be produced in which the wind turbines are superimposed upon photographs taken from the defined viewpoints.

Other factors also affect the visual appearance of a wind farm. Larger turbines rotate more slowly than smaller ones and a wind farm of fewer larger turbines is usually preferable to a wind farm of many smaller ones. In some surroundings, a regular area or straight line may be preferable compared to an irregular layout.

8.3.2 Noise from operating wind turbines

In densely populated countries, noise can sometimes be a limiting factor for the generating capacity that can be installed on any particular site. The noise produced by operating turbines has been significantly reduced in recent years by turbine manufacturers, but is still a constraint. This is for two main reasons:

1. Unlike most other generating technologies, wind turbines are often located in rural areas, where background noise levels can be very low, especially overnight. In fact the critical times are when wind speed is at the lower end of the turbine operating range, because then the wind-induced background noise is lowest.

2. The main noise sources (blade tips, the trailing edge of the outer part of the blade, the gearbox and generator) are elevated and so are not screened by topography or obstacles.

Turbine manufacturers may provide noise characteristic certificates, based on measurements by independent test organisations to agreed standards. The internationally recognised standard that is typically referred to is 'Wind turbine generator systems – Acoustic noise measurement techniques' (IEC 61400 Part 11 of 2003).

Standard techniques, taking into account standard noise propagation models, are used to calculate the expected noise levels at critical locations, which are usually the nearest dwellings. The results are then compared with the acceptable levels, which are often defined in national legislation.

The internationally recognised standard for such calculations is 'Acoustics – Attenuation of sound during propagation outdoors, Part 2: General method of calculation' (ISO 9613-2).

Sometimes the permitting authorities will require the project to conform to noise limits, with penalties if it can be shown that the project does not comply. In turn the turbine manufacturer could provide a warranty for the noise produced by the turbines. The warranty may be backed up by agreed measurement techniques in case it is necessary to undertake noise tests on one or more turbines.

8.3.3 Turbine loads

A key element of the layout design is the minimum turbine spacing used. In order to ensure that the turbines are not being used outside their design conditions, the minimum acceptable turbine spacing should be obtained from the turbine supplier and adhered to.

The appropriate spacing for turbines is strongly dependent on the nature of the terrain and the wind rose for a site. If turbines are spaced closer than five rotor diameters (5D) in a frequent wind direction, it is likely that unacceptably high wake losses will result. For areas with predominantly unidirectional wind roses, such as the San Gorgonio Pass in California, or bidirectional wind roses, such as Galicia in Spain, greater distances between turbines in the prevailing wind direction and tighter spacing perpendicular to the prevailing wind direction will prove to be more productive.

Tight spacing means that turbines are more affected by turbulence from the wakes of upstream turbines. This will create high mechanical loads and requires approval by the turbine supplier if warranty arrangements are not to be affected. Separately from the issue of turbine spacing, turbine loads are also affected by:

1. 'Natural' turbulence caused by obstructions, topography, surface roughness and thermal effects.

2. Extreme winds.

Defining reliable values for these parameters, for all turbine locations on the site, may be difficult. Lack of knowledge is likely to lead to conservative assumptions and conservative design.

Within the wind industry there is an expectation that all commercial wind turbines will be subject to independent certification in accordance with established standards or rules. A project-specific verification of the suitability of the certification for the proposed site should be carried out, taking into account the turbine design specifications and the expected climatic conditions of the site.

8.4   Infrastructure of wind farm

The wind farm infrastructure consists of:

8.4.1   Civil works

1. Roads and drainage.
2. Wind turbine foundations.
3. Met mast foundations (and occasionally also the met masts).
4. Buildings housing electrical switchgear, SCADA central equipment and possibly spares and maintenance facilities.

8.4.2   Electrical works

1. Equipment at the point of connection (POC), whether owned by the wind farm or by the electricity network operator.
2. Underground cable networks and/or overhead lines, forming radial 'feeder' circuits to strings of wind turbines.
3. Electrical switchgear for protection and disconnection of the feeder circuits.

4. Transformers and switchgear associated with individual turbines (although this is now commonly located within the turbine and is supplied by the turbine supplier).

5. Reactive compensation equipment, if necessary.

6. Earth (grounding) electrodes and systems.

8.4.3   Supervisory control and data acquisition (SCADA) system

1. Central computer.

2. Signal cables to each turbine and met mast.

3. Wind speed and other meteorological transducers on met masts.

4. Electrical transducers at or close to the POC.

The civil and electrical works, often referred to as the 'balance of plant' (BOP), are often designed and installed by a contractor or contractors separate from the turbine supplier. The turbine supplier usually provides the SCADA system. As discussed above, the major influence on the economic success of a wind farm is the energy production, which is principally determined by the wind regime at the chosen site, the wind farm layout and the choice of wind turbine. However, the wind farm infrastructure is also significant, for the following reasons:

1. The infrastructure constitutes a significant part of the overall project cost. A typical cost breakdown is given in Fig. 8.1.

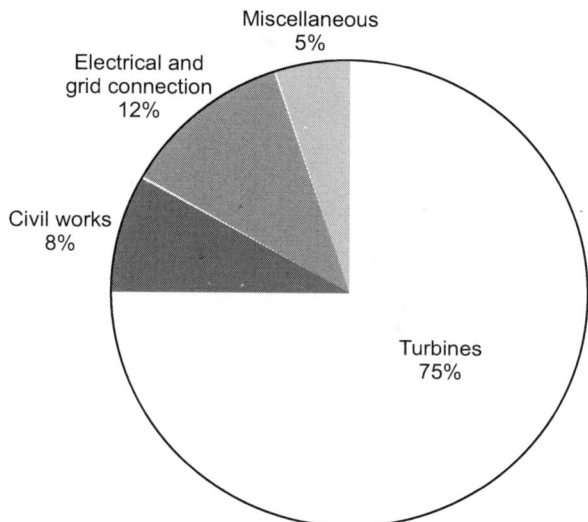

Figure 8.1: Typical cost breakdown for an onshore wind farm.

2. The civil works present significant risks to the project costs and programme. It is not unknown for major delays and cost over runs to be caused by poor understanding of ground conditions, or the difficulties of working on sites that, by definition, are exposed to the weather and may have difficult access.
3. The major electrical items (transformers, switchgear) have long lead times.
4. The grid connection works may present a significant risk to the programme. It is likely that works will need to be undertaken by the electricity network operator and the programme for these works is effectively out of the control of the wind farm developer. It is very unusual for electricity network operators to accept liability for any delay.

8.4.4 Civil works

The foundations must be adequate to support the turbine under extreme loads. Normally the design load condition for the foundations is the extreme, 'once in 50 years' wind speed. In Europe this wind speed is characterised by the 'three-second gust'. This is a sitespecific parameter, which will normally be determined as part of the wind speed measurements and energy production assessment for the site. For most sites this will lie between 45 and 70 m/s. At the lower end of this range it is likely that the maximum operational loads will be higher than the loads generated by the extreme gust and will therefore govern the foundation design.

The first step towards the proper design of the foundations is therefore the specification of a load. The turbine supplier will normally provide a complete specification of the foundation loads as part of a tender package. As the turbine will typically be provided with reference to a generic certification class, these loads may also be defined with reference to the generic classes, rather than site-specific load cases.

Although extremely important, the foundation design process is a relatively simple civil engineering task. A typical foundation will be perhaps 13 m across a hexagonal form and might be 1–2 m deep. It will be made from reinforced concrete cast into an excavated hole. The construction time for such a foundation, from beginning to end, can easily be less than a week.

The site roads fall well within normal civil engineering practice, provided the nature of the terrain and the weather are adequately dealt with. For wind farms sited on peat or bogs, it is necessary to ensure that the roads, foundations and drainage do not adversely affect the hydrology of the peat.

The wind farm may also need civil works for a control building to house electrical switchgear, the SCADA central computer, welfare facilities for maintenance staff and spare parts. There may also be an outdoor electricity substation, which requires foundations for transformers, switchgear and other equipment. None of this should present unusual difficulties. For upland sites, it is often beneficial to locate the control building and substation in a sheltered location. This also reduces visual impact.

8.4.5 Electrical works

The turbines are interconnected by a medium voltage (MV) electrical network, in the range 10 to 35 kV. In most cases this network consists of underground cables, but in some locations and some countries overhead lines on wooden poles are adopted. This is cheaper but creates greater visual influence. Overhead wooden pole lines can also restrict the movement and use of cranes.

The turbine generator voltage is normally classed as 'low', in other words below 1000 V and is often 690 V. Some larger turbines use a higher generator voltage, around 3 kV, but this is not high enough for economical direct interconnection to other turbines. Therefore, it is necessary for each turbine to have a transformer to step up to MV, with associated MV switchgear. This equipment can be located outside the base of each turbine. In some countries these are termed 'padmount transformers'.

Depending on the permitting authorities and local electricity legislation, it may be necessary to enclose the equipment within GRP or concrete enclosures. These can be installed over the transformers, or supplied as prefabricated units complete with transformers and switchgear. However, many turbines now include a transformer as part of the turbine supply. In these cases the terminal voltage of the turbine will be at MV, in the range 10 to 35 kV and can connect directly to the MV wind farm network without the need for any external equipment. The MV electrical network takes the power to a central point (or several points, for a large wind farm). A typical layout is shown in Fig. 8.2. In this case the central point is also a transformer substation, where the voltage is stepped up again to high voltage (HV, typically 100 to 150 kV) for connection to the existing electricity network. For small wind farms (up to approximately 30 MW), connection to the local MV network may be possible, in which case no substation transformers are necessary.

The MV electrical network consists of radial 'feeders'. Unlike industrial power networks, there is no economic justification for providing ring arrangements. Therefore a fault in a cable or at a turbine transformer will result in all turbines on that feeder being disconnected by switchgear at the substation. If the fault takes considerable time to repair, it may be possible to reconfigure the feeder to allow all turbines between the substation and the fault to be reconnected. Figure 8.2 shows two possible locations for the POC. Definitions of the POC vary from country to country (and are variously called delivery point, point of interconnection or similar), but the definitions are similar: it is the point at

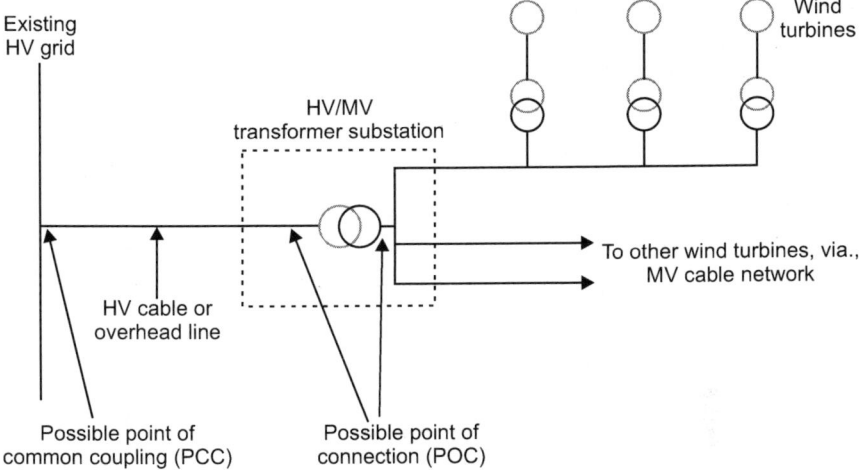

Figure 8.2: A typical electrical layout.

which responsibility for ownership and operation of the electrical system passes from the wind farm to the electricity network operator. More complex division of responsibilities is possible (for example, the wind farm developer may build and install equipment, which then is taken over by the network operator), but this is unusual.

The revenue meters for the wind farm will usually be located at or close to the POC. In some cases, where the POC is at HV, the meters may be located on the MV system to save costs. In this case it is usual to agree correction factors to account for electrical losses in the HV/MV transformer.

Figure 8.2 also shows a possible location of the Point of Common Coupling (PCC). This is the point at which other customers are (or could be) connected. It is therefore the point at which the effect of the wind farm on the electricity network should be determined. These effects include voltage step changes, voltage flicker and harmonic currents.

The design requirements for the wind farm electrical system can be categorised as follows:

1. It must meet local electrical safety requirements and be capable of being operated safely.

2. It should achieve an optimum balance between capital cost, operating costs (principally the electrical losses) and reliability.

3. It must ensure that the wind farm satisfies the technical requirements of the electricity network operator.

4. It must ensure that the electrical requirements of the turbines are met.

8.4.6   SCADA and instruments

A vital element of the wind farm is the SCADA system. This system acts as a 'nerve centre' for the project. It connects the individual turbines, the substation and meteorological stations to a central computer. This computer and the associated communication system allow the operator to supervise the behaviour of all the wind turbines and also the wind farm as a whole. It keeps a record on a ten-minute basis of all the activity and allows the operator to determine what corrective action, if any, needs to be taken. It also records energy output, availability and error signals, which acts as a basis for any warranty calculations and claims. The SCADA system also has to implement any requirements in the connection agreement to control reactive power production, to contribute to network voltage or frequency control, or to limit power output in response to instructions from the network operator. The SCADA computer communicates with the turbines, via., a communications network, which almost always uses optical fibres. Often the fibre-optic cables are installed by the electrical contractor, then tested and terminated by the SCADA supplier.

The SCADA system is usually provided by the turbine supplier, for contractual simplicity. There is also a market for SCADA systems from independent suppliers. The major advantages of this route are claimed to be:

1. Identical data reporting and analysis formats, irrespective of turbine type, this is important for wind farm owners or operators who have projects using different wind turbines.
2. Transparency of calculation of availability and other possible warranty issues.

In addition to the essential equipment needed for a functioning wind farm, it is also advisable, if the project size can warrant the investment, to erect some permanent meteorological instrumentation on met masts. This equipment allows the performance of the wind farm to be carefully monitored and understood. If the wind farm is not performing according to its budget, it will be important to determine whether this is due to poor mechanical performance or less-thanexpected wind resource. In the absence of good quality wind data on the site, it will not be possible to make this determination. Large wind farms therefore usually contain one or more permanent meteorological masts, which are installed at the same time as the wind farm.

8.4.7   Construction issues in wind farm

A wind farm may be a single machine or it may be a large number of machines, possibly many hundreds. The design approach and the construction method will, however, be almost identical whatever the size of project envisaged. The record of the wind industry in the construction of wind farms is generally good. Few wind farms are delivered either late or over budget. Newcomers to the wind industry tend to think of a wind farm as a power station. There are, however, some important differences between these two types of power generation. A conventional power station is one large machine, which will not generate power until it is complete. It will often need a substantial and complicated civil structure and construction risk will be an important part of the project assessment. However, the construction of a wind farm is more akin to the purchase of a fleet of trucks than it is to the construction of a power station. The turbines will be purchased at a fixed cost agreed in advance and a delivery schedule will be established exactly as it would be for a fleet of trucks. In a similar way the electrical infrastructure can be specified well in advance, again probably at a fixed price.

There may be some variable costs associated with the civil works, but this cost variation will be very small compared to the cost of the project as a whole. The construction time is also very short compared to a conventional power plant. A 10 MW wind farm can easily be built within a couple of months. To minimise cost and environmental effects, it is common to source material for roads from on-site quarries or 'borrow pits', where suitable. It may be necessary to seek permission for this from the permitting authorities.

8.4.8   Costs of wind farm

Wind farm costs are largely determined by two factors: the complexity of the site and the likely extreme loads. The site may be considered complex if the ground conditions are difficult – hard rock or very wet or boggy ground, for example – or if access is a problem. A very windy site with high extreme loads will result in a more expensive civil infrastructure as well as a higher specification for the turbines.

The cost of the grid connection may also be important. Grid connection costs are affected by:

1. Distance to a suitable network connection point.
2. The voltage level of the existing network.
3. The network operator's principles for charging for connections and for the use of the electricity system.

8.4.9   Commissioning, operation and maintenance

Once construction is completed, commissioning will begin. The definition of 'commissioning' is not standardised, but generally covers all activities after all components of the wind turbine are installed. Commissioning of an individual turbine can take little more than two days with experienced staff. Commissioning tests will usually involve standard electrical tests for the electrical infrastructure as well as the turbine and inspection of routine civil engineering quality records. Careful testing at this stage is vital if a good quality wind farm is to be delivered and maintained.

The long-term availability of a commercial wind turbine is usually in excess of 97%. This value means that for 97% of the time, the turbine will be available to work if there is adequate wind. This value is superior to values quoted for conventional power stations. It will usually take a period of some six months for the wind farm to reach full, mature, commercial operation and hence, during that period, the availability will increase from a level of about 80–90% after commissioning to the long-term level of 97% or more.

It is normal practice for the supplier of the wind farm to provide a warranty for between two and five years. This warranty will often cover lost revenue, including downtime to correct faults and a test of the power curve of the turbine. If the power curve is found to be defective, then reimbursement will be made through the payment of liquidated damages. For modern wind farms, there is rarely any problem in meeting the warranted power curves, but availability, particularly for new models, can be lower than expected in the early years of operation. During the first year of operation of a turbine some 'teething' problems are usually experienced. For a new model this effect is more marked. As model use increases, these problems are resolved and availability rises. After commissioning, the wind farm will be handed over to the operations and maintenance crew. A typical crew will consist of two people for every 20 to 30 wind turbines in a wind farm. For smaller wind farms there may not be a dedicated O&M crew but arrangements will be made for regular visits from a regional team. Typical routine maintenance time for a modern wind turbine is 40 hr per year. Non-routine maintenance may be of a similar order. There is now much commercial experience with modern wind turbines and high levels of availability are regularly achieved. Third party operations companies are well established in all of the major markets and it is likely that this element of the industry will develop very much along the lines associated with other rotating plant and mechanical/electrical equipment.

The building permits obtained in order to allow the construction of the wind farm may have some ongoing environmental reporting requirements, for example the monitoring of noise, avian activity, or other flora or fauna issues. Similarly there may, depending on the local regulations, be regulatory duties to perform in connection with the local electricity network operator.

Therefore, in addition to the obvious operations and maintenance activity, there is often a management role to perform in parallel. Many wind farms are funded through project finance and hence regular reporting activities to the lenders will also be required.

Major failures in wind turbines

9.1 Introduction

Reliability of wind turbines is a pre-requisite to ensure the healthy growth of wind energy. Even if new designs and prototypes performed by manufacturers and validated by certification bodies offer safer and more reliable wind turbines, their development and related improvement are based on the experience with turbines smaller than those currently being erected. Therefore the technology is still coming up against its limitations. To this end it has been recognised that there is a need for the continuous monitoring of major wind turbine components such as gear box, generator and rotor blades.

These components are seen to require substantial maintenance and repair efforts or even retrofits. Hence, periodic inspection of these components by any independent third party to ensure the safe and efficient operation is also necessary. The wind turbines are designed for a life span of about 20 years. However, numerous of studies have shown that some equipments in the nacelle interior as such the electrical system, hydraulic systems or drive train present very high failure rates, requiring so frequently repairs or replacement.

9.2 Causes of wind turbine failures

The root causes of these failures are diverse. In principle, the root causes can be subdivided into two causes, external factors, such as icing, lightning or storm and internal factors, as for example the failure of components or of the control system. Two third of all wind turbine plant failures lead eventually to a downtime of the plant. In one third of the downtime case, the plant can be taken into operation within a short-term, sometimes after only half day. In other cases the downtime lasts longer, particularly, when there is a requirement for repairs or replacements.

Wind turbine, which are equipped with a lightning protection can usually be taken into operation after a short-time. The most failures, accounting for more than 71%, are caused by internal factors. It is remarkable that the components defect, which mostly corresponds to technical components in the nacelle, accounts for almost 40% of the failure causes. The failures of the technical components can be considered as severe, since those mostly cause a plant downtime and require immediately repair or replacement.

The wind turbine failures have different impacts on the wind turbine operation.

9.3 Mechanical failures within the nacelle

9.3.1 Micro-pitting

The phenomenon micro-pitting can be found on several components, such as on the bearing, main shaft and high-speed shaft and specially on the gears. Micro-pitting also known as grey staining or frosting is a wear and tear phenomenon, which occurred on the surface areas of heavily loaded metallic component. By observing with unaided eyes, the affected area displays a frosted or dull grey appearance hence the names grey staining or frosting. The cause of this appearance can be recognised under high magnification of the affected area, where a large number of very small pores and micro-cracking can be observed.

Micro-pitting can be caused by many factors, although roughness and lubricant selection are the main sources. Generally the components surface, for example the teeth of the gears, are separated by a layer of a lubrication oil called Elastohyfrodynamic (EHD). Due to high temperature, excess load, high speed or water, the required oil (EHD) viscosity can decrease, rising so the probability of micro-pitting. The severest consequence of micro-pitting is the alteration of the components shape through the wearing. Since, the load is not evenly distributed over the surface and concentrates just on a small area. This can damage the component, for example the gears when they move through the mesh. Furthermore, micro-pitting can cause vibrations, noise, misalignment as also severe fatigue failure.

The essential measurement to avoid micro-pitting is to select the proper oil lubricant. Thereby, the oil viscosity plays a significant role. Moreover, it should be consider the operation conditions, as such temperature or speed as these have an significantly affect on the viscosity. The optimisation of geometry and metallurgy of the component can also minimise the chance of micro-pitting.

9.3.2 Corrosion

Corrosion is a chemical reaction or electrochemical reaction between a material and its environment, which can cause a remarkable alteration of its surface shape leading often to serious consequences. This phenomenon occurs usually on metallic material. Corrosion affects several components of the wind turbine, including the main shaft, high-speed shaft, generator, gearbox, as well the tower, which is often partly made of steel. However, specially affected from corrosion are offshore wind turbines, as the salt in the water and air increases the conductivity making the material more reactive. Offshore wind turbines have therefore additionally, the challenge to avoid the corrosion issues on the base of the turbine, which is placed under water and as well on the tower, which is partially sprayed with seawater.

The currently main applied methods to avoid the corrosion of wind turbine components are: The surface coating, which usually consist of a cooper-nickel alloy (CuNi 90/10) and makes the components material resistant against corrosion, the dehumidification systems and heating system. The dehumidification system has the purpose to reduce the humidity level of the air. As soon as the humidity level decreases to a certain value, the salt is physically unable to absorb enough moisture to initiate the corrosion process. This measurement also decreases the possibility of an electric short-circuiting.

The heating system has also as aim the reduction of the relative humidity. Although compared to the other methods, the heating system can be relative cost-intensive, making it a less attractive option.

9.3.3   Misalignment

Misalignment is common problem found in the wind turbine drivetrain. Misalignment issues affect the main components within the nacelle, including the generator, gearbox, main shaft or high-speed shaft. It is also remarkable that all these components belong to the rotating machinery group, where generally misalignments occur. In order to rotate freely and avoid any additional unwanted forces to the system, which can lead to serious damage of the equipments, the rotating machinery have to be proper aligned to each other. The process of alignment is nothing less than making the components co-linear under normal operating condition. On the other hand some components, require a certain defined misalignment, to allow an effective lubrication when operating.

The measurement of the alignment condition under normal operating condition is crucial as it can modify when the machine is operating. This can be caused by many factors, such as thermal growth, piping strain, foundation movement, machine torque and so on. Shaft alignments, for example, are measured when the machine is cold or in other words out of operation, consequently the value measured are not necessarily the alignment condition of the machine.

9.4   Inspections within the nacelle

In order to avoid all those mentioned failures and consequently downtimes and costs, it is necessary to perform regularly inspections. As already mentioned, due to an early, effective and accurate inspection, the maintenances operation, repairs and replacement can be exact planned in advance, reducing so the downtime period. The inspection requirements are different from component to component.

The Table 9.1 provides besides the inspection tasks per component also the inspection intervals according to the size of the plant.

Table 9.1: Wind turbine inspection task and intervals.

Component	Inspection task	Inspection interval		
		Annually	Every 2 years	Every 4 years
Hub	Cracks, corrosion, coating	>1.5 MW	0.3–<1.5 MW	<0.3 MW
Main shaft	Cracks, corrosion, coating, alignment	>1.5 MW	0.3–<1.5 MW	<0.3 MW
Screw fitting shaft-hub	Cracks, corrosion, clamping torques	>1.5 MW	0.3–<1.5 MW	<0.3 MW
Axel journals	Cracks, coating	>1.5 MW	0.3–<1.5 MW	<0.3 MW
Rotor bearing	Noise, sealing, lubrication, lubrication tank, lightning protection, shaft nut	>1.5 MW	0.3–<1.5 MW	<0.3 MW
Gearbox	Noise, visual inspection control of: abrasion, gearing, micro-pitting, wear debris analysis	>1.5 MW	0.3–<1.5MW	<0.3 MW
Drivetrain	Vibration analysis: constant measurement and capture of vibration data on different points of the drivetrain	>1.5 MW	0.3–<1.5MW	<0.3 MW
Lubrication oil supply	Condition monitoring, function and temperature difference measurement. Visual inspection of: oil level, foaming, filter contamination, debris analysis, function test of fluid pump	>1.5 MW	0.3–<1.5 MW	<0.3 MW

(Cont'd...)

Component	Inspection task	Inspection interval		
		Annually	*Every 2 years*	*Every 4 years*
Coupling and brake	Visual inspection control of: condition monitoring, abrasion, alignment	>1.5 MW	0.3–<1.5 MW	<0.3 MW
Torque arm	Motion, alignment, condition monitoring of rubber bearing	>1.5 MW	0.3–<1.5 MW	<0.3 MW
Generator (drivetrain with gearbox)	Bearing noise, sealing, fixing on bed plate, vibration, motor connection box, alignment	>1.5 MW	0.3–<1.5 MW	<0.3 MW
Generator (gearless)	Corrosion, sealing, inspection of crack on supporting body, air gap, bolted assembly	>1.5 MW	0.3–<1.5 MW	<0.3 MW
Temperature condition	Inspection of maximum bearing and lubrication oil temperature	>1.5 MW	0.3–<1.5 MW	<0.3 MW

The wind turbine inspections state-of-art technology relies mostly on visual and vibration analysis inspection. Generally, the key issues of wind turbine inspections are following:

1. The wind turbine has to be shut down for the inspection operations, which consequently results in a cost increase.
2. Repeatedly requirement of technicians climbing the wind turbine, specially offshores, sometimes under extreme weather condition, which besides causing more downtime also increases the risk of personal injury.
3. Lack of effective and accurate inspection.

The key challenge is to find inspection methods to avoid those issues. Inspection or condition monitoring is nothing less than the periodic view of the machine operating condition. Basically, the inspection methods can be divided in off-line and online inspection. An off-line inspection means that the wind turbine has to be shut down for the inspection period and/or requires the attention of a operator. On the opposite an online inspection can be carried out while the wind turbine is operating and usually the measurements devices are permanently integrated in the wind turbine system. Since the downtimes of wind turbine are very expensive, the interest in automated remote inspection methods is becoming more popular.

9.4.1 Vibration inspection

The vibration analysis inspection delivers several informations about the machine condition. It can help for example, to detect if the condition of the machine has changed, diagnose the cause of the change and classify its condition. Besides that, this method has the advantage to allow the inspection to be performed when the machine is running, avoiding so additional downtime. By definition vibration means the movement of a point, due to internal or external excitation, oscillating about a fixed reference point. It can be classified in: friction vibration (e.g., vibration from a bearing to a bearing house), free body vibration (e.g., shaft vibration), meshing and passing vibration (e.g., gear mesh vibration and blade pass vibration).

Basically, all those vibration types can be found in the wind turbine drivetrain. The frequency along the drivetrain varies from component to component, as it consists of several different components and each component has a particularly frequency.

On-site vibration monitoring

The on-site vibration measurement and monitoring is usually carry out by a compact portable device equipped with metering sensors.

Remote vibration monitoring

The remote vibration measurement is carried out by devices, which are installed in the wind turbine system and consists usually of connection box and several sensors, which are installed on different components of the wind turbine.

9.5 Major failures in the wind turbine components and its impacts

A wind turbine is a complex machine functioning in a complex environment. Wind turbines are built by the integration of various technologies and elements coming from aeronautics, mechanical engineering, hydraulics, electrical and electronic engineering, automation, informatics as well as civil works for the foundations. Figure 9.1 shows a quick review on the cause of failures occurred on a wind turbine. As for any integrated system, some of the components are more important than others, so, for a wind turbine, neuralgic components hence identified as critical are the gearbox, generator and rotor blades.

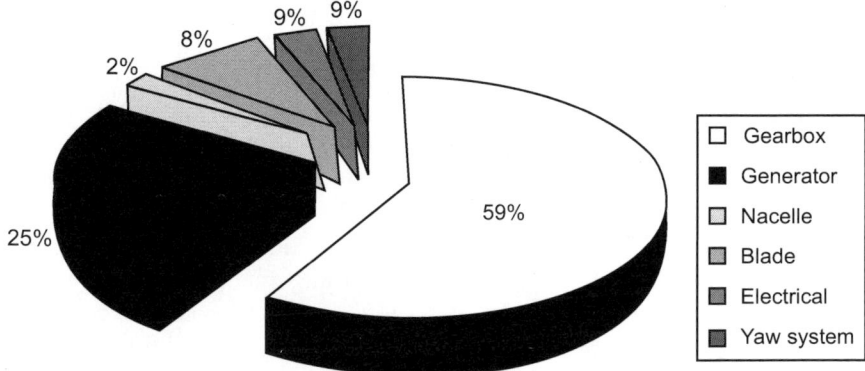

Figure 9.1: Wind turbine failures.

The various failures that occur in the wind turbines, frequency of the failure and the impact of those failures in the wind turbine operation. Thus, the failures which occurs in the neuralgic components such as gearbox, generator are having more down time and also it causes more economic losses for the wind farm operator.

These critical components have to be carefully controlled by means of a maintenance programme and regular inspections. If a fault is occurring on one of these main components, it may lead to an unsafe situation for the wind turbine itself and for others but inescapably lead to a financial risk with a loss of production. Therefore, a wind turbine has to be efficiently controlled by frequent inspections in order to protect the goods and preserve the continuity of production or, at least, to minimise as far as possible the shutdown period when performing maintenance work, repair or exchanging a component.

9.5.1   Gearbox

The conversion of the greatly differing rotational speeds of the rotor and the electric generator has given the designers of the first wind turbines many headaches. This situation has changed with the progress which has been made in gearbox technology.

Today, high-performance gearboxes with gear ratios of up to 1:100 and more are available. Regardless of this favourable situation, the gearbox has been and still is a source of failures and defects in many wind turbines because of their complexity and multiple moving parts which are possible weaknesses (e.g., bearings). The reasons for these failures are varied and can be attributed to manufacturing quality lapses when controlling the material supply and production activities, inappropriate design decisions or extreme weather situations on a given site.

The percentage of various failures occurs in the different parts of the gearbox. Statistically, in a wind turbine the gearbox is replaced every 5 to 7 years. Even in a year when a turbine gearbox has not failed, gearboxes represent single-handedly a financial risk because of their environmental impact through their acoustic emission which may lead to possible recourse action or financial resources to be implemented for maintenance work. Hence they require more maintenance as well as careful inspections.

The most common failure in the gearbox is bearing failures and the various reasons are heavily loading/lightly loading, misalignment, Incorrect package of thermal effects and poor lubrication. The other failures which occur in the gearbox are cracks/breakage in the gear teeth and the surface fractures. The reasons for these types of failures are excessive loading and due to vibration.

9.5.2   Generator

The Generator is the component of the wind turbine that transforms mechanical energy in to electrical energy and it is a very critical component of the wind turbine with high failure rates. The main reason is that the generator works with a power source supplying highly fluctuating mechanical power.

Two main categories of failures are known regarding the generators:

1. Winding failure which may result from defective insulation systems or poor winding design.
2. Mechanical failure because of early bearing fatigue which may result from poor lubrication.

The quality of the insulation is affected over the years because of functional stress. This stress reduces the electrical resistivity of insulations which creates an increase of the leakage currents and then can lead to incidents which may affect the safety of goods and persons as well as causing production losses due to shut down periods.

9.5.3   Rotor blades

The rotor blades of a wind turbine catch the energy of the wind. This energy is transformed into mechanical energy through the rotor that turns the main shaft of the wind turbine and then the generator to produce electrical energy. The blades are permanently stressed by environmental conditions like rain, moisture, temperature change, ice, UV radiation and of course lightning. There are several types of damages according to the localisation of the deviations:

Inside the blade: Cracks at the bonding resin, missing adhesive, discontinuities on the sandwich, delamination's within the Glass Fibre Reinforced Plastics (GFRP) or the sandwich, crack on web, excess of bonding resin, problems in the bonding, waves, air inclusions, etc.

Outside, on the surface: Erosion impact on the blade surface, deviations to observing laminate (spalling, flaking and cavities), deficient bond at the bonding surfaces, cracks and of course lightning strikes. It is obvious that rotor blades are a highly stressed part of a wind turbine because of the constant wind contact. They need regular inspection to evaluate their structural safety by experienced experts. Failures, damages and debris on the rotor blades can reduce the overall productivity of a turbine which can result in expensive repairs and important losses of production. Critical areas of a blade are beam and webs, bonding (web, leading edge and trailing edge) and the sandwich at the maximum profile depth (Table 9.2).

Table 9.2: Possible damages on blade and time for correction

Possible damages found on blade	Recommended time for correction
Minor superficial defects found inside and outside	24 months
First signs of damage in the structure, several surface defects and thin cracks at the bonding	12–24 months
Damage in the structure, cracks in the shell and in the bonding, several surface defects	3–12 months
Major defects in the main structure and important cracks that decrease the aerodynamic structure	0–3 months with possible reduction of power
Damage that cannot comply with a safe operation	Stop of the production for safety reason

Regarding the economic facts and costs to exchange a blade, it makes sense to be sure that the blades are in a good condition and that they will last a long time. To face this operational reality, it is strongly recommended to take care of the blades by performing yearly blade inspections as well as making sure that repairs and cleaning are part of the preventative maintenance programme implemented by the manufacturer or other maintenance service company mandated by the owner or its operator.

9.5.4   Wind turbine inspection

The periodic inspections of the wind turbine and the main components are carried out by an independent technical expert who shall have access to the relevant technical documentation. On the basis of these documents, a specific checklist with the evaluation criteria is prepared in order to perform the inspection. The assessment and results have to be based on the guideline for the certification of wind turbines of Germanischer Lloyd and the IEC standards in their latest editions The inspection plan includes all critical components related to the power production, the protection system and the protective measures for safety of the staff.

The inspection report is written and signed by the technical expert. At least, this report has to contain the following information:

1. Manufacturer, type and serial number of the wind turbine and the tower.
2. Location of the owner and the operator of the wind turbine.
3. Operating hours and total energy produced.
4. Date and weather condition on the day of inspection.
5. Persons present at the inspection.
6. Detailed description of the scope of inspection
7. Remarks, defects and deviations found. A timeframe shall be prescribed for repairs and recommended corrective actions.
8. Result of the inspection.

The owner has to file the inspection reports for the operating life of the wind turbine.

9.5.5 Various types of wind turbine inspections

At appropriate times for periodic technical inspections during life time of the wind farm/wind turbine with specific inspections on main components are discussed below.

Fitted inspections at appropriate times

1. At the workshop as soon as nacelles and blades have successfully passed the manufacturing quality control process.
2. At the site during the delivery of the components from unloading until the definitive storage condition.
3. At certain stages during assembly/construction of the wind turbine or wind farm.

Periodic inspections

The complete turbine shall be examined closely by visual inspection on individual components as well as rotor blades. The structural integrity of the wind turbine, including machinery, functioning of the safety and braking systems, shall be checked at the period stated below:

1. Just after the commissioning performed by the manufacturer to inspect the work done and to establish a snag list before signing the official takeover certificate of each wind turbine.
2. 6 months prior to the contract completion (turbine sale contract or operation and maintenance contract) in order to address to the manufacturer the remaining issues before the end of the warranty period.
3. Before a merge and acquisition of the wind farm.

4. Whenever during the life time of wind farms/turbines, recommended every 2 or 4 years.

Specific inspections on main components

Gearbox inspection: In order to perform a general assessment of the gearbox's condition, investigations are aided by a visual and by a video-endoscopic inspection inside the gearbox to detect wear and deviations on bearings and gears. The possibility of inspecting the main drive by vibration sensors analysis is also available. A gearbox inspection scope includes these following issues:

1. Oil condition sample and analysis.
2. Gearbox condition.
3. Gearbox's cooling system and oil lines of the cooling systems.
4. Inspection of the gearbox, via., a flexible video-endoscope on all visible bearings, cogs, pinions and sprockets.
5. Inspection of the drive train components, via., off-line vibration sensor analysis allows detecting the evidence of early fault defects especially at the gearbox and generator.

Generator inspection: A generator inspection work consists of these following investigations:

1. Verification high speed shaft coupling, shaft alignment and calculation of the compensations to realign the coupling train.
2. Check if abnormal noise is heard from the bearings.
3. Generator's cooling system.
4. Insulation test of the generator (stator and rotor).

Blades inspection: Both interior and exterior sides of each blade and its root region are checked for air inclusions, cracks, lightning damages, erosion wears, delamination, defective bonds and other quality problems. The condition of the joint seals, lightning protection systems and additional aerodynamic parts are checked as well as existing rain deflectors or possible repaired areas. The lightning protection system can be verified with a measurement of the ohmic resistance.

To briefly sum up the situation, on the one hand, there is the demand for wind energy which is rapidly expanding and on the other hand, the heart of the industrial process that converts wind energy into electrical energy consists of a complex machine made of critical and neuralgic components that may lead to important production losses and repair costs. If we draw a parallel between these observations, wind turbine inspection is an essential service that facilitates our wind industry to produce reliable wind power and for enhanced safety related aspects.

Improving wind turbine performance using nanomaterials

10.1   Introduction

The wind is free, commonly available and can provide clean, pollution-free energy. Today's wind-turbines are very high tech but the life time of wind turbines is less due to moisture, efficiency and friction losses. This chapter present new design scheme of light weight structure for wind turbine tower. This design scheme is based on the integration of the nano-structured materials. The objective is to accomplish the weight reduction by optimising the wall thickness of the tower while combining the appropriate material properties into optimisation. By using new techniques the efficiency, life time of a wind turbine can be enhanced.

Wind power has been around for a fairly long time. Wind turbines simply convert kinetic energy into mechanical energy. Similar to solar energy, wind power is very clean since it produces no harmful emissions. The concept of converting wind energy into electrical energy is very simple and in a way, similar to any form of traditional generation methods, using a source to power a generator but without the harmful emissions. Wind turbines can be thought of as the exact opposite as what fans does, instead of using electricity to generate wind, they use wind to generate electricity. The blades on the wind turbine convert the kinetic energy from the wind into mechanical energy that is then used to turn a shaft in a generator to generate electricity. This is similar to traditional generation techniques but instead of using coal or oil to generate steam to power a generator, wind is used directly.

10.2   Components of wind turbines

Anemometer: Measures the wind speed and transmits wind speed data to the controller.

Blades: Most turbines have either two or three blades. Wind blowing over the blades causes the blades to 'lift' and rotate.

Brake: A disc brake, which can be applied mechanically, electrically, or hydraulically to stop the rotor in emergencies.

Controller: The controller starts up the machine at wind speeds of about 8 to 16 miles per hour (mph) and shuts off the machine at about 55 mph. Turbines do not operate at wind speeds above about 55 mph because they might be damaged by the high winds.

Gear box: Wind turbines rotate typically between 40 rpm and 400 rpm. Generators typically rotate at 1200 to 1800 rpm. Most wind turbines require a step-up gear-box for efficient generator operation (electricity production). Gears connect the low-speed shaft to the high-speed shaft and increase the rotational speeds from about 40 to 60 rotations per minute (rpm) to about 1000 to 1800 rpm, the rotational speed required by most generators to produce electricity. The gear box is a costly (and heavy) part of the wind turbine and engineers are exploring 'direct-drive' generators that operate at lower rotational speeds and don't need gear boxes.

Generator: Usually an off-the-shelf induction generator that produces 60-cycle AC electricity.

High-speed shaft: Drives the generator.

Low-speed shaft: The rotor turns the low-speed shaft at about 30 to 60 rotations per minute.

Nacelle: The nacelle sits atop the tower and contains the gear box, low- and high-speed shafts, generator, controller and brake. Some nacelles are large enough for a helicopter to land on.

Pitch: Blades are turned, or pitched, out of the wind to control the rotor speed and keep the rotor from turning in winds that are too high or too low to produce electricity.

Rotor: The blades and the hub together are called the rotor.

Tower: Towers are made from tubular steel, concrete, or steel lattice. Because wind speed increases with height, taller towers enable turbines to capture more energy and generate more electricity.

Wind direction: This is an 'upwind' turbine, so-called because it operates facing into the wind. Other turbines are designed to run 'downwind,' facing away from the wind.

Wind vane: Measures wind direction and communicates with the yaw drive to orient the turbine properly with respect to the wind.

Yaw drive: Upwind turbines face into the wind, the yaw drive is used to keep the rotor facing into the wind as the wind direction changes. Downwind turbines don't require a yaw drive, the wind blows the rotor downwind.

10.3 Issues and challenges of wind turbines

There are four types of issues present in wind turbines:

1. Distribution problem.
2. Variation in wind speed.

3. Power control.
4. Life, weight, power losses and efficiency.

10.4   Reasons for failure of wind turbines

There are many reasons for wind turbine failure, such as:
1. 45% of cost breakdown, mainly in the turbine nacelle.
2. Electrical control system 13%.
3. Gearbox 12%.
4. Yaw system 8%.
5. Generator 5%.
6. Hydraulics 5%.
7. Grid connections 5%.

10.5   Proposed solution

Nanotechnology is solving challenges in wind power component supply. The wind turbines life time can be increased by using nano paints and weight can be reduced by using fibre glass and efficiency can be increased by coating. Individuals are usually assured that the electricity they are using is actually produced from a green energy source that they control. Once the system is paid for, the owner of a renewable energy system will be producing his own renewable electricity for essentially no cost and can sell the excess to the local utility at a profit.

10.6   Power control in wind turbines

There are many types of losses in wind turbine such as:
1. Mechanical losses.
2. Stray load losses.
3. Iron losses.
4. Copper losses.

Since a single Wind Turbines Generator (WTG) has limited capacity, much less than a conventional power generator, a wind power plant is called wind farm.

Normally a wind farm consists of many WTGs connected together by overhead lines or cables. Their power output is collected and transmitted to the grid through an alternating current (AC) or direct current (DC) line, after voltage stepup at the substation in the WPP. Some WPPs now have a capacity comparable to that of conventional power generators.

10.7　Proposed solution to other issues

Wind turbine parameters need to improve are shown in Fig. 10.1.

1. De-icing coatings.
2. Self-cleaning coatings.
3. Weight saving.
4. Lubricants.
5. New sealants.
6. Power pack improvements.

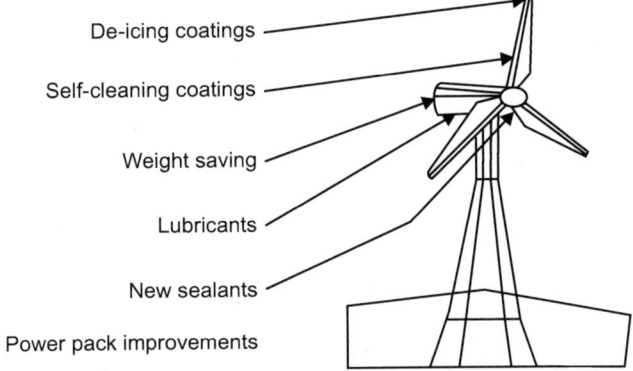

Figure 10.1: Improving wind turbine parameters.

10.8　Power control

The speed of turbine should to directly proportional to the turbine output power. So, we need to improve wind turbine efficiency and reduce losses to improve power coefficients. Figure 10.2 shows the turbine power and turbine speed versus wind speed.

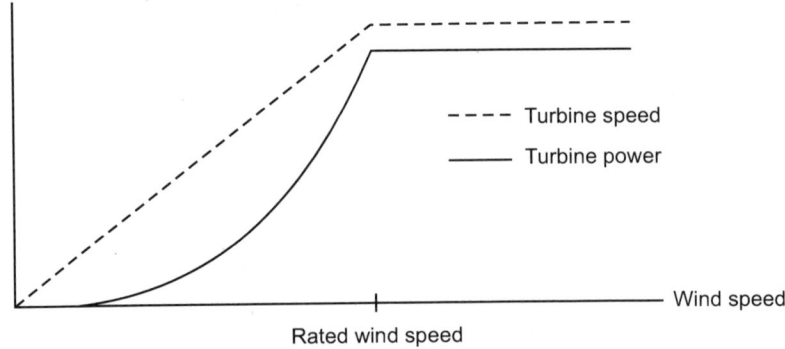

Figure 10.2: The turbine power and turbine speed versus wind speed.

For calculating the TSR the following equation can be used

$$TSR = \frac{\text{Tangental speed at the blade tip}}{\text{Actual wind speed}} = \frac{R_\omega}{V_0} \qquad \dots(10.1)$$

10.9   Methodology

1. The turbine efficiency can be increased, which raises the energy production a few per cent.
2. The noise emission at low wind speeds can be reduced.
3. Variable-speed systems also allow torque control of the generator and therefore the mechanical stresses in the drive train can be reduced.
4. Resonances in the turbine and drive train can also be damped and the power output can be kept smoother.
5. By lowering the mechanical stress the variable-speed system allows a lighter design of the wind turbine.
6. The economical benefits of this are very difficult to estimate but they may be rather large. Table 10.1 shows solution to above issues.

Table 10.1: Proposed solution to above issues.

Wind power problem	Possible solution	Example
Ice-build up on blades and sensors	Non wettable surface, treatment: Degussa micro-porosity of fibreglass which reduce porosity to prevent ice build up	Lotus plant: Non-wettable self cleaning leaves due to nanostructured rough surface with wax crystals
Dirt build up on blades	Self cleaning surfaces, TiO_2 nano coating	Pilkington's self cleaning glass
Damage to blades	Use protective coating, e.g., non-scratch surfaces	Automotive, e.g., Mercedes
Strength/weight of composites	Nanocomposite improve properties	Automotive use, e.g., Toyota chevrolet
Reliability of rotating machine and replacing worn out components	Nano lubricant for improved wear resistance at all temperatures and pressures	Nanolub contains particles which behaves like mini ball bearings
Hydraulic system leaks	Novel sealants based on Nanocomposite	Nanocor: Brake systems, pitch and yaw control, on board cranes, locking system, pumps, drivers, oil tanks, filters, pressure valves and control system
Start up and orientation requires grid power	Carbon nanotubes as fuel storage	Hydrogen storage for fuel cells using nanotubes fuel storage

(Cont'd...)

Wind power problem	Possible solution	Example
Control system	SCADA for systems for grid operators to condition monitoring, remote control and full monitoring and mini generators and energy storage for start up and nil wind	Industry SCADA system

10.10   Proposed nano materials and techniques

The Table 10.2 shows the method of improvements for wind turbines.

Table 10.2: Proposed method.

Strength/weight improvements	Nanoparticles in composites	Nano materials
Tensile strength up 40% Tensile modulus (elasticity) up 68%	Improved mechanical properties Better diffusion barrier characteristics	Carbon nanotubes 50–110 stronger than steel 1/6th the weight
Flexural strength up 60%	Better processability	Used in body armour, yacht masts, tennis
Flexural modulus (bending) up 120%	Better transparency	racquets and car body panels
	Electrical conductivity	The best electrical and
Distortion temperature from 65% to 152%	Better burning behaviour	thermal conductors
Improved flame retardant properties	Better surface properties	Result in plastic that conducts electricity and heat Storage of hydrogen

10.11   Impact of nanotechnology on wind turbines for clean energy

The impact of nanotechnologies in various area of turbine is given below:

1. For wind turbines through by using composite materials based on carbon nanotubes the light weight and high strength rotor blades can be developed.

2. Carbon nanotubes composites can be used in wind turbines which provide excellent conductivity.

3. Opportunities to improve: nanotechnology is solving challenges in wind power component supply.

4. The wind turbines life time can be increased by using nano paints and weight can be reduced by using fibre glass and efficiency can be increased by coating.

5. Nano scale Nondestructive Testing (NDT) composites.
6. Visualising damage by a nano scale interferometric technique.
7. Damage by a nano scale interferometric technique using lasers.
8. Strain data can determine damage and residual life.
9. Used for periodic inspection of turbine blades.

To sum up, by taking into account all the analysis it can be concluded that the best solution is to use more than one technology together such as PV, wind turbines, fuel cells and geothermal. The wind turbines efficiency and life time efficiency can be increase by using nano materials and techniques in future. The wind turbines provide clean energy and use of clean energy in future is only solution for clean healthy environment.

Composite material and carbon nanotubes in wind turbine blades

11.1 Introduction

Developing sectors, increasing population and rising energy demand are the reasons for improvements in energy generation fields. So to meet the needs of this power demand, new technologies are being invented to be optimised as renewable energy sources. Wind energy is one such source of energy available among many which has been used for power generation. Wind turbines generate electricity as they convert the mechanical energy of the turbines driven from wind. But many times failure of turbine blades takes place due to various reasons such as high load, seasonal impacts, creep formation in the blades, high stress, lackof proper maintenance, mechanical breakdown, gearbox issues and even human error. Prevenient turbines were made up of single material and it was observed that they had very short life and very poor failure resistance. Subsequently, new types of materials were invented by blending some dissimilar materials to form composites. Today, turbines are made by these composite materials which are far more expedient.

Despite of manufacturer's best intention and so called assurance or guarantees, wind turbines will fail. There are many factors which have to be taken under consideration like design, material, plant setup, area etc. Focusing on how the different composites used to make turbine blades, to reduce wear and improve failure resistance, prospect of all rigorous conditions is the dominant objective of this chapter. By varying composition of the material different tests for the material testing were carried out at different temperatures. Tensile strength, stiffness, rigidity and coefficient of rigidity were calculated under different loadings such as tensile-tensile loading, compression-compression loading, tensile-compression loading with repetitive cycles. Different composites were tested and their properties were compared with each other in order to find out the best composite material amongst them. After tests it was found out, that carbon fibre composites give the best result but they are very expensive and it is not economical to use carbon fibres every time. Later on it was observed that composite known as basalt fibre composite has the nearly equal strength to that of the carbon fibre composite besides that they are easily available and cheap. So these basalt fibre composites can possibly replace carbon fibre composites.

11.2　Composite materials

Composite materials are the combination of two materials which results in new material having better properties than individual materials. Composites are the materials made up of two materials at microscopic scale having two different phases. One has to clearly understand the difference between alloys and composites. In alloys the materials used conserve their mechanical, physical and chemical properties. Reinforcement and matrix are main parts of composite. Reinforcements are generally fibres which add many new properties in the material. Matrix is generally polymer or ceramics or metal. Constituent materials differ in properties from each other.

They are heterogeneous and homogeneous at microscopic and macroscopic level respectively. As per the requirements, respective materials can be combined to add appropriate properties in the initial material, this significantly alters the characteristics of that material. Composites are dominantly used in order to add and enhance desired properties of the primitive material.

11.3　Classification of composites

Classification of composite materials is shown in Fig. 11.1.

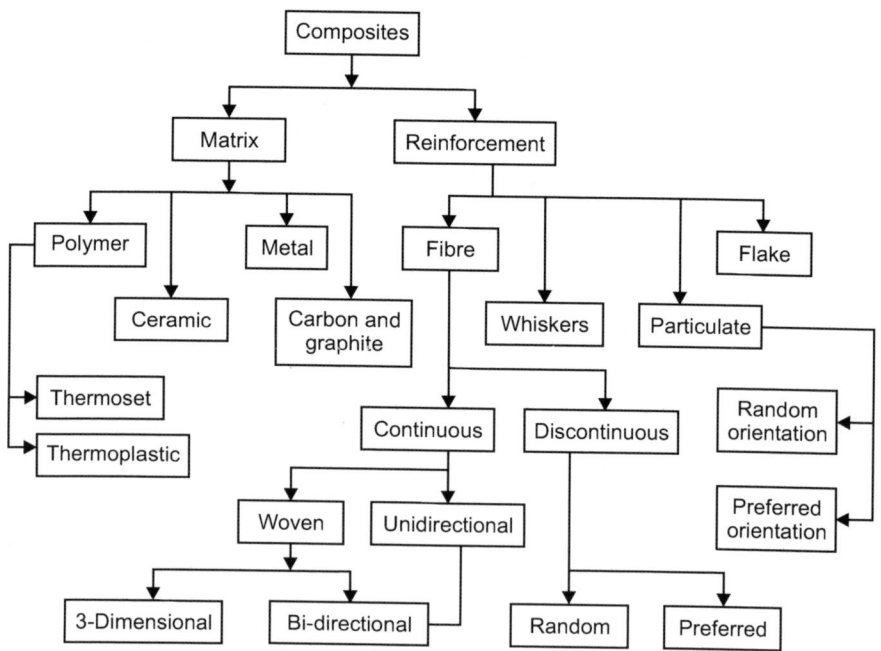

Figure 11.1: Classification of composite materials.

11.4   Wind turbine rotor blades: Construction, loads and requirements

Among all the parts of wind turbines (blades, hub, gearbox, generator, nacelle, tower), composite materials are used in blades and nacelles. The main requirements to nacelles, which provide weather protection for the components, are the low weight, strength and corrosion resistance. Typically, nacelles are made from glass fibre composites.

Blades represent the most important composite based part of a wind turbine, whose properties quite often determine the performances and lifetime of the turbine. In fact, a rotor is the highest cost component of a wind turbine. Still, the failure rates of wind turbine blades are of the order of 20% within three years and this is surely too much. Increasing the reliability and lifetime of wind blades is an important problem for the developers of wind turbines.

The wind turbine blades are built as follows. A blade consists of two faces (on the suction side and the pressure side), joined together and stiffened either by one or several integral (shear) webs linking the upper and lower parts of the blade shell or by a box beam (box spar with shell fairings). The box beam inside the blade is adhesively joined to the shell. Figure 11.2 shows the section of the blade of the wind turbine.

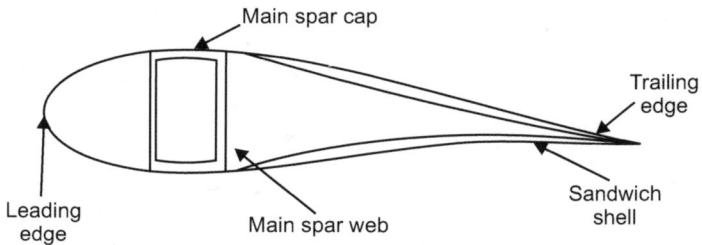

Figure 11.2: Section of the blade of the wind turbine.

Wind turbine blades are subject to the external loading, which includes the flapwise and edgewise bending loads, gravitational loads, inertia forces, loads due to pitch acceleration, as well as torsional loading. The flapwise load is caused mainly by the wind pressure, while the edgewise load is caused both by gravitational forces and torque load. The biggest edgewise bending moment is at the blade root.

The flapwise and edgewise bending loads cause high longitudinal, tensile and compressive stresses in the material. The upwind side of the blades is subject to tensile stresses, while the downwind side is subject to compression. The flapwise and edgewise bending moments lead to the fatigue damage growth. These two moments are responsible for 97% of the damage in blades.

The wind blades are also subject to cyclic loadings, caused by wind variations, turbulences, wind shear and other effects, like pressure variations of air around the tower. During the functioning, the stability of the blade shape (aerodynamic profile), high durability and reliability should be ensured. The shape stability corresponds to the minimum deflection of the blade under wind loads.

This is achieved by increasing the moment of inertia of the blade (using the corresponding blade design) and by increasing the flexural stiffness of the wind blade material. The flapwise bending is resisted by the spar, internal webs or spar cabs inside the blade, while the edges of the profile carry the edgewise bending. Table 11.1 summarises the roles of the parts of wind blades in maintaining the blade shapes.

Table 11.1: Wind blade parts and their functions in maintaining the blade shapes.

Part	Function	Materials used
Blade shell	Maintaining the blade shape, resisting the wind and gravitational forces	Strong, lightweight composites
Unsupported parts of the shell	Resisting the buckling load	Thickened sandwich structures with light core materials and multidirectional face laminates
Integral web, spars or box beam	Resisting the shell buckling/ shear stresses due to flapwise bending	Biaxial lay-ups at ±45°
Adhesive layers between composite plies and the web and the blade shell	Ensuring the out-of plane strength and stiffness of the blade	Strong and highly adhesive matrix

The wind blades are produced with multiaxial fabrics. Often, ±45° laminates are used in the blade skin and in the shear web. In the root area, triaxial materials are utilised, ±45°/90°. Spar caps are produced from unidirectional composites, with some biaxial plies.

In unsupported parts of the wind shell, the sandwich composites are used. They ensure the shape stability of blade shell. The core of the sandwich bears the shear load, while the composite skin resists the bending stresses. The sandwich structures ensure much higher stiffness than the monolithic composites. The sandwich core materials, placed between two composite plies, are typically polymer foams, balsa wood or in some cases honeycomb types (nomex).

The main requirements to wind turbine blades can be summarised as follows:

1. High strength (to withstand even extreme winds, as well as gravity load).
2. High fatigue resistance and reliability (to ensure the stable functioning for more than 20 years and 108 cycles).

3. Low weight (to reduce the load on the tower and the effect of gravitational forces).

4. High stiffness to ensure the stability of the aerodynamically optimal shape and orientation of the blade during the work time, as well as clearance between blade and the tower.

11.5 Composites for wind turbine blades: Main constituents and manufacturing

In order to ensure the required shape stability, strength and damage resistance of the wind turbine rotor blades, the blades are produced from long fibre reinforced polymer laminates. In these composites, long fibres ensure longitudinal stiffness and strength, while the resin matrix is responsible for fracture toughness, delamination strength and out-of-plane strength and stiffness of the composite.

11.5.1 Fibres

The stiffness of composites is determined by the stiffness of fibres and their volume content. Most often, E-glass (i.e., borosilicate glass called 'electric glass' or 'E-glass' for its high electric resistance) fibres are used as main reinforcement in the composites.

The main properties of E-glass fibres are as: Young modulus E 70–77 GPa, density 2.55–2.64 kg/m^3, diameter 8–15 μm, failure strain 4.5–4.9%.

With increasing the volume content of fibres in UD composites, the stiffness, tensile and compression strength increase proportionally, yet, at high volume content of fibres (after 65%), there might be dry areas without resin between fibres and the fatigue strength of the composite reduces. Typically, the glass/epoxy composites for wind blades contain up to 75 weight % glass.

Many investigations toward the development of fibres, which are stronger than the usual E-glass fibres, have been carried out. The high strength fibres (which are still used seldom in practice, but represent a promising source of the composite materials improvement) include glass fibres with modified compositions (S-glass, R-glass, etc.), carbon fibres, basalt and aramid fibres.

S-glass (i.e., high strength glass, S means 'Strength' here) developed in the 1960s for military applications, has 40% higher tensile and flexural strengths and 10–20% higher compressive strength and flexural modulus, than the E-glass. The main properties of S-glass are: Young modulus E 86–90 GPa, density 2.46–2.49 kg/m^3, failure strain 5.4–5.8%. Still, the S-glass is much more expensive than E-glass. S2 glass was developed in the 1968 by Owens Corning company as a commercial, non-military version of S-glass. S-glass and S2 glass fibres have the same composition (magnesium alumino-silicate).

The main differences are in sizing and certification procedure. The price of S2-glass is around 10 times of that of E-glass. R-Glass fibres, introduced by
Vetrotex in 1968, are produced with a calcium aluminosilicate glass with less silica and added oxides. The main properties of R-glass are: Young modulus E 84–86 GPa, density 2.55 kg/m^3, failure strain 4.8%. Some other special glasses developed by Owens Corning are ECRGLAS and Advantex. Relatively recently, in 2006, Owens Corning company developed WindStrandTM glass fibres, which have 15% higher stiffness and up to 30% higher strength when compared to traditional glass fibre reinforcements and show very good fatigue properties under both tension and compression loading.

Carbon fibres attracted large interest of industry and research community as a very promising alternative to the glass fibres. Carbon fibres have much higher stiffness and lower density than the glass fibres (Young modulus E = 220–240 GPa, density 1.7–1.8 kg/m^3, failure strain 0.7%), thus, allowing the thinner blade profile as well as stiffer and lighter blades. However, they have relatively low damage tolerance, compressive strength and ultimate strain and are much more expensive than the E glass fibres. Furthermore, carbon fibre reinforced composites are very sensitive to the fibre misalignment and waviness: even small misalignments lead to the strong reduction of compressive and fatigue strength. In some cases, the problem of efficient wetting carbon fibres in vacuum infusion has been observed, thus, leading to the use of more expensive prepreg technology for the producing carbon fibre based composites.

The carbon fibres are used mainly by the wind turbine makers Vestas and Gamesa in structural spar caps of large blades. Another possibility is to use carbon layers locally in the root end of wind blades, with some kind of transition from carbon to the glass reinforcements in other parts of the blade. As noted by Fecko, while carbon fibres provide the highest stiffness among all the fibres, high strength glasses provide the best combination of stiffness, strength and impact resistance. Further alternative options of non-glass, high strength fibres include aramid and basalt fibres.

Aramid (aromatic polyamide) fibres have high mechanical strength and are tough and damage tolerant. However, the compressive strength of aramid fibres is even lower than that of carbon. Their mechanical properties are as follows: Young modulus E 133–135 GPa, density 1.44 kg/m^3, failure strain 2.5–4 %. Aramid fibres were introduced by the company DuPont as Kevlar. However, aramid fibres absorb moisture, can degrade under the influence of ultraviolet radiation and have relatively low adhesion to polymer resins. Basalt fibres, with their good mechanical properties, represent also an interesting alternative to the glass fibres. Basalt fibres are 30% stronger, 15–20% stiffer and 8–10% lighter than E-glass and cheaper than the carbon fibres. In some cases, natural fibres can be used as well. A French company La Tôlerie Plastique (LTP) was awarded the JEC 2010 Innovation Award for wind energy for the development of small wind turbines made of 100% biodegradable materials, consisting of flax fabrics with a PLA matrix.

The interest to hybrid reinforcements (E-glass/carbon, E-glass/aramid, etc.) has been growing during the last decade as well. According to the analysis of costs and benefits of replacement of glass fibres by carbon fibres for a given 8 m blades carried out by Ong and Tsai, the full replacement would lead to 80% weight savings and cost increase by 150%, while a partial (30%) replacement would lead to only 90% cost increase and 50% weight reduction.

11.5.2 Matrix

Due to the low weight requirement to the wind blades, polymers are the main choice as the matrix material for the wind blade composites. As noted above, matrix of composite controls fracture toughness, delamination strength and out-of-plane strength and stiffness of the composite and influences the fatigue life of the composites. Typically, thermosets (epoxies, polyesters, vinylesters) or (more seldom) thermoplastics are used as matrixes in wind blade composites.

Thermosets based composites represent around 80% of the market of reinforced polymers. The advantages of thermosets are the possibility of room or low temperature cure and lower viscosity (thus, allowing better impregnation and adhesion). Initially, polyester resins were used for composite blades. With the development of large and extra-large wind turbines, epoxy resins replaced polyester and are now used most often as matrixes of wind blade composites. While polyester is less expensive and easier to process (needs no post-curing), epoxy systems are stronger (high tensile and flexural strength) and more durable as compared with polyester resins. Epoxy matrixes ensure better fatigue properties of the composites. The production of epoxy based composites is more environmentally friendly. Still, recent studies (e.g., by Swiss company DSM Composite Resins) support arguments for the return to unsaturated polyester resins, among them, faster cycle time and improved energy efficiency in the production, stating that the newly developed polyesters meet all the strength and durability requirements for large wind blades.

Thermoplastics represent an interesting alternative to the thermoset matrixes. The important advantage of thermoplastic composites is their recyclability. Their disadvantages are the necessity of high processing temperatures (causing the increased energy consumption and possibly influencing fibre properties) and, difficulties to manufacture large (over 2 m) and thick (over 5 mm) parts, due to the much higher viscosity. The melt viscosity of thermoplastic matrices is of the order 10^2–10^3 Pax s, while that for thermosetting matrix is around 0.1–10 Pax s. Thermoplastics (as differed from thermosets) have melting temperatures lower than their decomposition temperatures, and, thus, can be reshaped upon melting. While the fracture toughness of thermoplastics is higher than that of thermosets, fatigue behaviour of thermoplastics is generally not as good as thermosets, both with carbon or glass fibres. Other advantages of thermoplastics include the larger elongation at fracture, possibility of automatic processing and unlimited shell life of raw materials.

Further, the development of matrix materials which cure faster and at lower temperatures is an important research area. Resins with faster cure and lower curing temperature allow reducing the processing time and automating the manufacturing. In some cases, thixotropic agents, like fumed silica and certain clays, are used to control viscosities of resins during manufacturing.

In several works, the possibilities of improvement of composites properties by adding nanoreinforcement in matrix were demonstrated. Additions of small amount (at the level of 0.5 weight %) of nanoreinforcement (carbon nanotubes or nanoclay) in the polymer matrix of composites, fibre sizing or interlaminar layers can allow to increase the fatigue resistance, shear or compressive strength as well as fracture toughness of the composites by 30–80%. Graphite particles fibre coatings on glass fibres allow increasing the fatigue life up to 100 times.

Summarising the brief overview, one can state that apart from the basic solution (epoxy/E-glass composites) widely used for medium and large wind blades, there are several very promising directions of development of stronger, more reliable and economically producible composites, among them:

1. High strength fibres (strong glasses, carbon, aramid, other fibres) can ensure higher stiffness and sometimes better strength and damage resistance of composites, the disadvantages are higher costs and in some cases lower compressive strength (carbon fibres) and high sensitivity to local defects (e.g., misalignment). Hybrid composites with mixed E-glass and high strength fibres allow to achieve the combination of higher stiffness (due to carbon fibres) with limited cost increase.

2. Thermoplastic resin matrixes, as opposed to the widely used epoxy matrix, are recyclable, can be reshaped upon melting and have higher fracture strain. Their disadvantages are high processing temperatures and higher viscosity, leading to the more expensive and difficult processing times. Resins with faster cure and lower curing temperature allow us to reduce the processing time and automate the manufacturing. Further, the nano-modified matrixes or sizing on fibres have a potential to increase the fatigue strength of the composites.

11.6   Properties of composite materials

1. Strength.
2. Toughness.

3. Corrosion resistance.
4. Wear resistance.
5. Stiffness.
6. Reduce weight.
7. Acoustic insulation.
8. Energy dissipation.
9. Fatigue life.
10. Conductivity.
11. Thermal resistance, etc.

Constituents of composite materials are called as reinforcements or matrix. Reinforcements are harder, discontinuous and stronger while matrix is continuous.

11.7   Disadvantages of composite materials

1. High cost.
2. Transverse properties are sometimes found to be weak.
3. Disposal and reuse of composites can be a difficult task.
4. Joining two composite parts is tough.
5. Maintenance and repair is difficult.
6. They are brittle, so there are chances of getting damaged.
7. Matrix has low toughness.
8. Manufacturing and using composites is a risk to health.

11.8   Carbon nanotubes in wind turbine blades

In recent years, wind power has become an increasingly attractive source of energy generation. Wind turbines from the leading manufacturers are generally guaranteed to operate over a lifetime of approximately 20 years. To extend this working lifetime and to meet increasing energy needs requires advanced wind blades that are stiff, strong and have extended fatigue resistance.

This section discusses epoxy and vinylester (VE) resins as attractive alternatives to the standard polyester for turbine blades. VE resins are inexpensive and have several other qualities, including strong chemical resistance, low viscosity during processing and mechanical strength. The properties of VE resins can be enhanced significantly by adding fillers such as CNTs. However, because of strong attractive interactions, CNTs tend to aggregate to form bundles, which in turn create a highly entangled network that prevents them from dispersing in the VE resin. The answer lies in dispersing individual CNTs in the polymer matrix and then stabilising them. Surface modification of CNTs, covalent functionalisation and non-covalent modification using small molecules and polymer dispersants can be used to overcome dispersion and stabilisation issues.

A sonication method is used to disperse multi-walled CNTs (MWCNTs) into a VE system and screened eight different dispersing agents. Samples are centrifuged to evaluate which dispersing agents could stabilise the CNT suspensions. According to this qualitative method, if a suspension does not show visible phase separation after centrifugation, the dispersing agent is assumed to be a suitable candidate for the stabilisation of MWCNTs. The resistance to sedimentation of 0.1wt% CNTs dispersed in VE resin after centrifugation. Neat VE resin did not give stable MWCNT dispersions. But the dispersing agent B60H, a polyvinyl butyral with low residual polyvinyl alcohol content, proved to be effective against sedimentation. Optical micrographs of the suspensions obtained after sonication. After 50 hr, we noticed that the MWCNTs were beginning to agglomerate. By contrast, the system containing B60H as dispersing agent did not exhibit agglomeration after such time. The influences of CNT and different dispersing agent concentrations on the mechanical properties of the system using standard tensile testing. The VE composites prepared with 0.1wt% MWCNTs and dispersing agent exhibited improved ductility and thus increased toughness when compared to neat VE (Fig. 11.3).

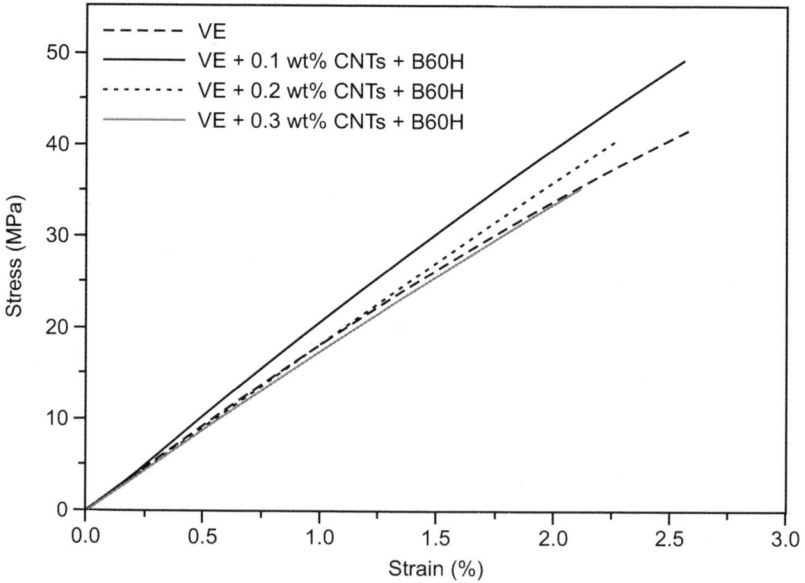

Figure 11.3: Stress-strain curves for neat VE and VE-based composites with CNTs and B60H. The B60H concentration was fixed at 1x that of the CNTs.

Increasing the concentrations of MWCNTs and B60H diminishes this reinforcement role. Thus, the use of the dispersing agent is beneficial in achieving improved tensile properties for low levels of CNTs in VE composites.

To sum up, it was found that B60H is a suitable dispersing agent for the long-term stable suspensions of MWCNTs in VE, preventing the nanotubes from agglomerating for at least 50 hr. The strategies for obtaining long-term stable MWCNT-resin dispersions in thermoset resins presented in this was studied may help to increase the potential use of CNTs in industrial applications. The next step is to prepare macroscopic assemblies of CNTs that mimic the intrinsic properties of individual tubes. Work is also needed to develop new, green techniques to modify the surfaces of the CNTs. Ideally, bio-based thermoset resins could be used as an alternative to current petroleum-based resins.

Role of lubricants in wind energy

12.1   Introduction

Lubricants is a substance (such as grease) capable of reducing friction, heat and wear when introduced as a film between solid surfaces. Lubricants can be divided into four major categories: (i) liquid lubricants, (ii) lubricant greases, (iii) solid lubricants and (iv) gaseous lubricants.

12.2   Types of lubricants

12.2.1   Liquid lubricants - lubricant oils

In this category, liquid lubricants can be divided on more sub categories: Vegetable and animal origin lubricants, mineral and synthetic based oils.

Vegetable and animal oils

The vegetable and animal lubricants were the first to be used by mankind. Due to their chemical inertia and due to the increasing demands on the lubricant specifications these were progressively abandoned and replaced by synthetic and petroleum based products. Although these were abandoned, their usage has seen an increase due to environmental reasons. As an example, their usage on Nordic countries where to lubricate equipment used on forestry activity. As its biggest weakness it can be said that these types of oils have rapid oxidation and show low resistance to high temperatures.

Mineral based oils

Petroleum based lubricants are mainly made of natural hydrocarbons resulting from organic residue decomposition and are usually referred to as mineral oils. These lubricants can be classified by the type of basis of oils.

Paraffin basis

These oils are characterised by its low specific weight, low freeze point, great oxidation resistance and low viscosity variation with temperature thus making its viscosity index high, (about 100). On one hand, paraffin based oils are usually elastomer friendly, (elastomers used to build seals and others). On the other hand, the high molecular weight of certain molecular chains present on this basis may carry the crystallisation point just below the ambient temperature.

Naphthenic basis

These oils are characterised by its high specific weight, high freezing point, little oxidation resistance and great viscosity variation with temperature making its viscosity index quite low, (about 50). Naphthenic based oils are relatively aggressive to the elastomers used on seals. These are greatly miscible, highly volatile and exhibit great fluidity at low temperatures.

Mineral oils are then designated as paraffin or naphthenic based depending on the predominance of one of the basis on its constitution. There are other products that are a mixture of both basis, these are largely used as lubricants because it allows for a combination of the following properties:

1. Availability on a wide range of viscosities.
2. Low volatility.
3. Good/high degradation resistance.
4. Creation of a good corrosion protection.
5. Low price.

Synthetic based oils

Synthetic based oils are lubricants synthesised from hydrocarbons or its constitutive elements, having as its basis petroleum derived products, vegetable oils, (this type of basis seems to have grown the most during the last years) and others. The knowledge on the hydrocarbons's basic structure and the understanding of the basic properties needed for lubrication have allowed for the development of different synthesis methods. These basis have been in development on the last decades with the goal of solving particularly difficult lubrication problems. Generally synthetic based oils have higher performance than the mineral based ones, specially on:

1. Oxidation resistance.
2. Viscosity index.
3. Coefficient of friction.

The advantages of using synthetic based oils are more expressive at high or low temperatures. The main disadvantage of these oils is just the much higher price when comparing to other solutions like the mineral based ones. There is a huge variety of these synthesised oils.

Synthetic hydrocarbons

Polyalphaolefins and benzenes are the most common types. Some of the most important characteristics of these lubricants are enumerated below:

1. Similar to the mineral hydrocarbons on its chemical structure.
2. Good compatibility with the elastomers usually used on seals.

3. Good miscibility with mineral oils.
4. Excellent thermal stability.
5. Great behaviour at low temperatures.
6. Need for anti-oxidation additives.
7. Food and pharmaceutical industry grade oils can be produced.

Polyglycols

One of the main characteristics of these lubricants is that they allow for a low coefficient of friction. This makes them specially attractive for applications where high sliding is expected (worm gears). Compatibility with seals and other polymers must be carefully analysed, specially if the working temperature is above 100°C. Generally very low miscibility with mineral oils is observed.

Esters

Ester based oils result from a reaction between acids and alcohols followed by water separation. There are several types of esters, each one of them with its own characteristics that will greatly influence the final lubricant properties.

High thermal resistance and great behaviour at low temperatures are generally observed. Some esters despite being biodegradable are fastly biodegradable. This factor seems to have increasing influence on the choice of a lubricant, since now-a-days people are more aware of the environmental questions. A good choice of an ester base can allow for a coefficient of friction on the same level of the polyglycols. One of the problems shown by ester based lubricants is the low hydrolytic stability. Hydrolytic stability depends not only on the ester type but also on the additive package used on the final blend of the lubricant.

Silicons

Some of the most important characteristics of these lubricants are enumerated below:

1. Chemically inert.
2. Fire resistant.
3. Non mixable with water.
4. Non toxic.
5. High oxidation resistance at high temperatures.
6. Good thermal stability.

12.2.2   Lubricant greases

Lubricant greases result from the dispersion of a thickening agent on a lubricant oil. Most of the lubricant greases have soap as thickener, but some organic products can be used as thickeners too. Lubricant greases are used when continuous oil lubrication is not viable and when protection against possible outside contaminant particles is required and outside contamination is to be avoided, like on pharmaceutical and food industries.

This type of lubricants have its properties limited by the thickening agent, base oil and additives.

Solid lubricants

A solid lubricant is a film of solid material constituted by organic or inorganic compounds, that is between the lubricating surfaces.

Inorganic solid lubricants

There are three different types of compounds used as inorganic solid lubricants.

1. Gelatinous solids: Materials like graphite and molybdenum disulphide are disposed in different layers. Inter atomic forces inside a layer are quite strong but the inter atomic forces on the interface between layers is quite weak allowing for easy sliding between the layers.

2. Soft solids mixtures: There is a huge variety of inorganic solids like lead, calcium oxide, talc, silver iodide and lead monoxide that are used as lubricants.

3. Surface protection by chemical reaction with the surface: There are countless compounds that can chemically react with the surface. Surface coatings like chlorides, oxides, phosphates and sulphides are well known.

Organic solid lubricants

This types of lubricants are usually divided in two categories.

1. Soaps, waxes and fat: On this category metallic soaps like calcium, sodium and lithium can be included. As for the waxes one can mention beeswax and fat acids.

2. Polymeric films: On this last category synthetic substances like Teflon (PTFE), can be included. One of the main advantages of this coatings is its great resistance to environmental aggressions.

12.2.3   Gaseous lubricants

Gaseous lubrication is in some aspects analogous to liquid lubrication, since the principles of hydrodynamic lubrication can be applied. Both are viscous fluids, but there are some fundamental differences that separate them. Gases have much lower viscosity than the liquids despite having much higher compressibility. The load carrying capacity of a gaseous lubricant is much lower than the one of a liquid lubricant.

12.3   Additives in lubricants

Adding chemical agents to the lubricant oils, (usually called additives), has the objective of giving the lubricant certain desired properties. Some additives give the lubricant new properties of great utility that were not initially available and others just improve its natural properties.

Additives were firstly used on lubricant oils around 1920 and since then it has been increasing over the years. Now-a-days practically almost every lubricant has at least one additive, but some have more than 5 types. The quantity of the used additive can vary from a few hundredths to 30%. Additive packages have been increasing the available oils properties , thus paying a big role on the developing of engines and all of the industrial machinery.

12.3.1   Viscosity index improving additives

Viscosity index improving additives are used in order increase the Viscosity Index of the base oil, namely lower the viscosity at low temperatures and/or increase the viscosity at high temperatures.

The viscosity index of a lubricant is improved by adding high molecular weight polymers.

The viscosity of a lubricant, that has been added with polymers, is modified to an higher extent at high temperatures than at lower temperature. So, if at a lubricant oil of low viscosity is added a viscosity index improving additive a small increase on its viscosity is observed at low temperatures, (the lubricant keeps its fluidity), but an higher relative viscosity increase is observed at high temperatures. This additives are used for engine oils, on fluids for automatic gearboxes and on hydraulic systems.

12.3.2   Anti wear and extreme pressure additives

Anti wear and extreme pressure additives are used to reduce the friction at extreme lubrication conditions. This products can be classified on three general categories:

1. Lubricity agents.
2. Anti wear additives.
3. Extreme pressure additives.

12.3.3   Lubricity agents

The lubricity agents are usually added to the base oils in order to reduce the friction on limit film lubrication conditions. It should be noted that the application should dictate the type and quantity of lubricity agents. This type of additives is used, for example, when phenomena like stick and slip is to be avoided.

12.3.4 Anti-wear (AW) additives

The anti-wear additives are used to increase the anti-wear properties of the base oil forming a protective film from the reaction with the metallic surfaces on contact.

12.3.5 Extreme pressure (EP) additives

These additives have the objective of avoiding the adhesion between metallic surfaces, when the conditions of extreme pressure and relative sliding are verified. The EP additives can chemically combine with the contacting metallic surfaces forming a protective layer. When the rupture of the lubricant film is verified, (due to high sliding speeds and/or high contact pressures), this thin layer will protect the surfaces from extremely severe failure phenomena.

EP additives are usually used on gears where the contact pressure between the teeth can be well above 700M Pa.

Now-a-days there is a huge variety of extreme pressure additives. Lubricant oils can be prepared in order to satisfy practically all of the operational demands on gear applications. When formulating these lubricants the oil thermal and chemical stability must be ensured, since the EP additives are prone to induce lubricant instability.

12.3.6 Antioxidants

These are used to avoid, modify or delay the reaction of the hydrocarbons with oxygen in order to avoid all of the harmful situations that oil oxidation can cause. Oil oxidation can generate acid compounds soluble on the lubricant. These oxide compounds can usually not only increase the oil viscosity, but also make it corrosive to some metals forming sludge and varnishes that can adhere to the mechanical organs.

An antioxidant additive can actuate in different manners. It can directly affect the lubricant oil, or it can affect the metallic surfaces that usually it is in contact with. In order to have the desired stability and corrosion resistance the base lubricant oil should be meticulously refined and its anti-corrosion additives should be carefully selected in order to have the best lubricant properties.

Other additives

Other types of additives are often used on a lubricant:

1. Detergents: These are used to avoid the formation of deposits of extremely viscous compounds on the lubricant.
2. Corrosion inhibitors: These are usually the same additives as Antioxidants, since oxidation can result on the formation of acid components that can deteriorate the surfaces.

3. Surface oxidation inhibitors: By physical or chemical interaction with certain metals, these form a continuous and extremely strong layer over the metallic surfaces that does not allow water to be in contact with those surfaces.

4. Foam inhibitors: The most used foam inhibitor is silicon. Usually foam inhibitors are extremely effective because only a few parts per million are enough to have the desired effect. These avoid foam formation caused by intense lubricant agitation during operation. It can be used on any type of liquid lubricant.

12.4 Physical properties of lubricanting oils

Physical properties of the lubricants are usually defined by the base lubricant.

12.4.1 Viscosity

From all of the physical and chemical properties that need to be considered on lubrication, viscosity is the most important. Viscosity of a fluid is defined as the resistance opposed by the fluid at all of the internal shear deformation. This opposing force can be calculated using the Newton's formula relative to the laminar flow between a mobile surface with a velocity V and a fixed surface.

12.5 Lubricanting oils specifications

Two types of specifications are available for the lubricant oils:
1. Viscosity specifications.
2. Service specifications.

12.5.1 Viscosity specifications

Viscosity specifications can be established with two ends:
1. Identification: There are refining or manufacturing specifications that take into account viscosity tolerances for certain ranges of viscosity.
2. Usage: These are imposed by the consumer and are function of the use given to the lubricant oils. There are certain ranges of maximum and minimum viscosity at certain temperatures.

These classifications are based only on the lubricant oil viscosity.

There are various professional societies that classify lubricants by viscosity range. Some examples of these societies are:
1. SAE   :   Society of Automotive Engineers,
2. ISO   :   International Standards Organisation,
3. AGMA :   American Gear Manufacturers Association,
4. ASTM :   American Society for Testing and Materials.

SAE Viscosity classification

This classification is used and almost universally accepted for the classification of automotive lubricant oils (with or without additives). It is exclusively based on the oil viscosity and it doesn't evaluate the oil quality it just gives the information about the its viscosity at a certain temperature.

A different classification is proposed depending on the oil application. For example the SAE J306 is the classification given to the lubricant oils for automotive gearboxes and differentials and the SAE J300 is for engines.

12.6 Characterisation of the wind turbine gear oils

Four fully formulated wind turbine gear oils were tested. Among them two Mineral, (MINE, MINR) a Polyalkylene Glycol, (PAGD) and a Poly-α-olefin (PAOR). All of these lubricants are marketed as high performance gear oils for application in wind turbine gearboxes and all of them are ISO VG 320 lubricant oils. The gear oils physical properties (viscosity and bulk density) need to be evaluated before any other test. This process consisted on the use of an Engler viscometer to determine the engler viscosity at three different temperatures in order to later on calculate the Vogel and ASTM D341 constants for each oil. The dynamic viscosity can also evaluated on a rotational viscometer. Various density measurements can be performed at various temperatures in order to assess its variation with temperature.

12.6.1 Viscosity

In order to have accurate values for the viscosity at a specified temperature, the Viscosity vs Temperature curve had to be determined for each oil. It is of great interest to have accurate values for the viscosity at a given temperature, since it will, not only, affect the way the implemented power loss model will respond, but also will help on the experimental results interpretation.

Engler viscometer

The Engler viscometer has a recipient where the oil to be evaluated is poured. There is a small hole at the bottom of that recipient where a wood pointer should be inserted or removed in order to stop or allow the oil to flow from it. To keep the oil at a selected temperature, this recipient is inside another recipient with a fluid between them (oil or water). This fluid is heated by a resistor thus allowing for temperature control. This two recipients are supported by a three legged support that allows adjustments in order to keep both recipients levelled. There are two thermometers, one on the tested oil and the other one on the heat carrying fluid. This allows one to control the temperature that is desired for the viscosity measurement.

Rheometer

In order to evaluate the character of the gear oils, (Newtonian or Non-Newtonian at atmospheric pressure) tests are performed on the available rheometer. The rheometer, is a coaxial measurement system according to Searle's principle. The measuring shaft is rotating inside the testing substance and it is powered by an electrical motor. The opposing torque is then measured and shown on the control instrument display. By knowing the opposing torque and the angular speed, the viscosity of the fluid can be determined. By performing tests at various shear strain rates the fluid behaviour (Newtonian or Non-Newtonian) can be tested.

12.6.2   Density

The evolution of the density with temperature at atmospheric pressure was performed for gear oils. The density is used to calculate the dynamic viscosity, the thermal expansion coefficient, etc.

The densitometer used, has two ways of evaluating the oil sample:

1. It can directly extract a 2 ml oil sample from a recipient.
2. The oil sample can be injected on the device with a special syringe.

The oil sample temperature should be within the range from 0 to 40°C. In order to obtain the density curve three values for the density at three different temperatures were registered and the thermal expansion coefficient was calculated. Figures 12.1a,b present the obtained curves for viscosity and density. Since it has an high percentage of additives using a vibro-viscometer a another viscosity measurement was performed at a constant temperature of 100°C in order to evaluate if there was a viscosity change over time. No changes were observed.

Table 12.1 shows the properties according to the manufacturers data sheets. The physical properties and chemical composition of the four gear oils were characterised and are shown in the Table 12.2. All wind turbine gear oils are ISO VG 320 grade, meaning they have a viscosity of 320cSt at 40°C. As shown in Table 12.2 the measurements show that these oils satisfy the ISO standard.

The measurements performed on the rheometer showed that all of the gear oils have a Newtonian behaviour at the tested conditions.

It should be observed that the MINR oil shows the highest viscosity variation with temperature between all wind turbine gear oils. PAGD oil shows a viscosity 52.76cSt at 100°C, which is quite high since it is more than twice the viscosity of the MINR oil at this temperature. MINE oil sits in between both and shows great improvements on its viscosity at higher temperatures, comparing with MINR. As for the density one should remark that PAGD is more dense than water at 40°C.

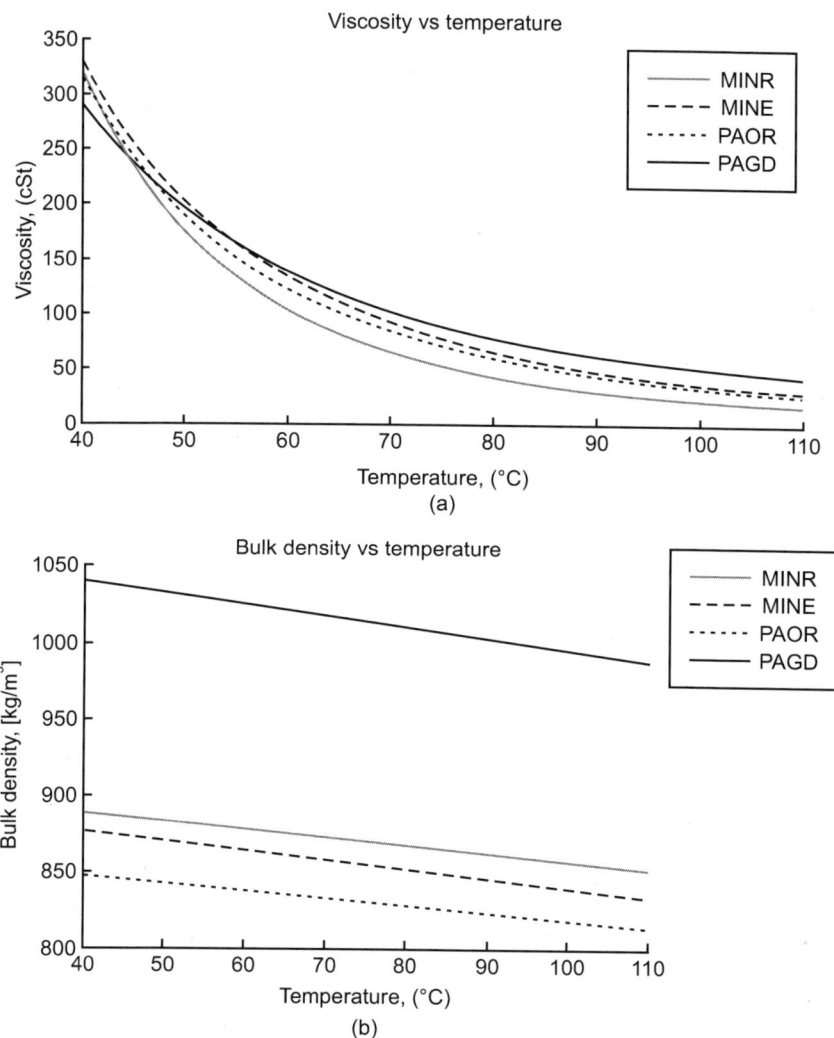

Figure 12.1: Variation of the wind turbine gear oils properties with temperature:
(a) viscosity and (b) bulk density. Viscosity curves calculated according to ASTM
D341 procedures, bulk density curves derived from measurements performed in
the range of 15 to 35°C.

12.7 Selecting oil for a gearbox

Selecting the right oil and care of the oil is one of the main objectives so that
the system can operate in optimal conditions throughout its lifespan. Selecting
the right oil should be the shared responsibility of the user, the manufacturer
of the machine, the manufacturer of the mechanical components (bearings, etc.),

Table 12.1: Physical properties of the wind turbine gear oils.

Parameter Base oil	Unit [–]	Standard	MINE Mineral	MINR Mineral	PAGD Polyalkylene glycol	PAOR Poly-α-olefin
Physical properties						
Density @ 15°C	[g/cm³]	ASTM D4052	n/a	0.904	1.051	n/a
Viscosity @ 40°C	[cSt]	ASTM D445/DIN 51562–1	320	320	320	320
Viscosity @ 100°C	[cSt]	ASTM D445/DIN 51562 - 1	n/a	24.03	54	35.5
Viscosity index	[–]	ASTM D2270/DIN ISO2909	n/a	95	237	158
Inflammability point	[°C]	ASTM D92/DIN ISO2592	n/a	210	102.33	210
Pour point	[°C]	ASTM D97/DIN ISO3016	n/a	–9	–39	–42
Corrosion resistance properties						
Copper corrosion, 3 hr at 100°C n/a		[–]	ASTM D130	n/a	1a	Pass
Rust, method A	[–]	ASTM D665	n/a	Pass	Pass	Pass
Rust, method B	[–]	ASTM D665	n/a	Pass	Pass	Pass
Wear properties						
FAG FE-8 roller wear	[mg]	DIN 51819-3	n/a	3	Pass	n/a
Micropitting test	[–]	FVA 54/7	n/a	GF>10	n/a	n/a
FZG A20/8.3/100°	[–]	DIN 51354	n/a	>12	n/a	>12

Table 12.2: Physical properties of the wind turbine gear oils.

Parameter	Unit	MINE	MINR	PAGD	PAOR
Chemical composition					
Zinc (Zn)	[ppm]	<1	0.9	1	3.5
Magnesium (Mg)	[ppm]	<1	0.9	1.4	0.5
Posphorus (P)	[ppm]	460	354.3	1100	415.9
Calcium (Ca)	[ppm]	2	2.5	0.8	0.5
Boron (B)	[ppm]	36	22.3	1.0	28.4
Sulphur (S)	[ppm]	6750	11200	362	5020
Physical properties					
Density @ 15°C	[g/cm^3]	0.893	0.902	1.059	0.859
Thermal expansion coefficient (α_t)	[K^{-1}]	-6.7×10^{-4}	-5.8×10^{-4}	-7.1×10^{-4}	-5.6×10^{-4}
Viscosity @ 40°C	[cSt]	328.59	319.25	290.26	324.38
Viscosity @ 70°C	[cSt]	92.72	65.87	102.33	87.92
Viscosity @ 100°C	[cSt]	37.88	22.41	51.06	35.27
VI	[−]	166	85	241	155
Vogel constants					
k	[−]	0.2033	0.0815	1.5068	0.1875
b	[−]	1072.492	1051.819	638.4922	1043.2304
c	[−]	105.173	87.129	81.3362	100.5561

the oil supplier and the supplier of the filters. The gearboxes of wind turbines are characterised by the low and high speeds and the alternative loads to which they are subjected, which gives rise to the use of oils with extreme pressure (EP) additives.

The base fluids used can be mineral or synthetic. The mineral fluids are products derived from petroleum, while the synthetic fluids are produced through synthesis. The synthetic oils can be: Polyalphaolefins (PAO), Ester oils (E) or Polyglycols (PAG).

The properties of a new oil for wind turbine gearboxes must be in accordance with German standard DIN 51517 Part 3 and with the following requirements:

Parameter	Methodology	Criteria
Viscosity index	ISO 2909	Minimum 90
Oxidation stability	ASTM-D2893-Amended	Increase in viscosity to 121°C < 6%
Corrosion of steel	ISO7120	Negative
Corrosion of copper	ISO2160	<1B

(Cont'd...)

Parameter	Methodology	Criteria
Foam	ASTM-D892	75/10 75/10 75/10
FZG Scuffing	ISO 14635-1	>=12
Micropitting	FVA 54	>=10
Filterability	AFNOR NF E48690	Pass
	5 Microns	
Cleaning	ISO 4406/99	16/14/11
FE 8	DIN 51819	<30 mgr/80 hr
Brugger	DIN 51347	> 50 N/mm^2
Air retained at 90°C	ASTM-D 3427	< 15 minutes
Weld load	ASTM-D 2783	>250 Kg
Wear 1800 rpm/20 kg/54°C/60min	< 0.35 mm	
Demusibility 82°C	ASTM-D1401	< 15 mins.

12.7.1 Problems with gearbox oils

The main problems associated with the lubricants detected in the gearboxes through predictive and proactive maintenance, which are as follows:

1. Micropitting.
2. Foam and retained air.
3. Remaining oil life.

Micropitting

Although micropitting is not a new phenomenon, it has not been given much significance until now. However, it is known to affect gear-tooth precision and, in many cases, it is the first type of fault. Micropitting is a surface fatigue phenomenon that occurs in Hertzian contacts, caused by cyclic contact stress and the plastic flow of asperities. This results in the formation of microcracks and micropitting and loss of material. Micropitting is also referred to as fatigue scoring, flecking, spalling, glazing, frosting, grey staining, microspalling, peeling, etc.

Micropitting is surface damage which occurs in high load-bearing systems and is characterised by the presence of small holes in the surface, revealing an inner surface with cracks. It first appears on the load-bearing surface and then extends towards the root (dedendum) of the tooth, the area where the gear works more.

All gears are susceptible to micropitting, including external, internal, spur, helical and bevel gears. Micropitting can occur in heat-treated materials, including nitrided, hardened, tempered, etc. It is still not fully understood why some oils are more prone to the formation of micropitting than others.

Micropitting results in the loss of tooth profile and could potentially lead to macropitting, breakage of the tooth, noise and vibration. According to Robert Errichello, it is clear that both micro and macropitting are fatigue processes, the only difference being that micropitting is much smaller. Metallographic analysis shows that the cracks that occur in micropitting and in macropitting have the same morphology, but are of very different size. In many cases micropitting is not harmful to the gear surface and its development may even be stopped when the tribological conditions of the system are re-established.

Micropitting can be sometimes be eliminated by polishing during the gear rolling process, when it is said that the gears have been 'cured'. The depth of micropitting is no greater than 10 microns and is difficult to see with the human eye, only becoming perceptible when measuring more than 40 microns.

Micropitting is also not considered to be a problem in itself, but it can sometimes lead to macropitting. This is a failure caused by surface roughness and not Hertzian contact stress. Greater roughness produces greater stress under the surface, which results in micropitting. The resistance of new oils to micropitting is evaluated using the high-speed gear test, for example, the FZG FVA54 I-IV micropitting test and the BGA micropitting test. There are few micropitting control methods currently in use. The main methods are visual inspections (borescope), oil analysis and destructive testing of gears.

Visual inspections

Visual inspections offer a good method of controlling micropitting in gears in service.

Appearance: The appearance is a change in the tone of the gear - dull, grey, etc. It is difficult to see.

Where to look: Micropitting starts on the load-bearing surface and is mainly due to asperities remaining after the manufacture of the gears. This is why this phenomenon usually appears at the beginning of the machine's life, during the first million gear cycles. Micropitting begins as a surface contact at the edges of the gears where there are crests, undulations, peaks, etc. It is usually accompanied by other failure modes such as scuffing, macropitting and abrasion. When the damage caused by micropitting varies from tooth-to-tooth it is mainly because of variations in tooth geometry or surface roughness. The micropitting pattern on a gear set may repeat at a particular frequency of a common factor. For example, a gear set with a 20/45 tooth combination may display similar micropitting on every fifth tooth.

Causes of micropitting

Micropitting occurs under elastohydrodynamic lubrication (EHL) where the thickness of the lubricant film is of the same order as the composite surface roughness and the load is borne by surface asperities and lubricant. Under elastohydrodynamic lubrication (EHL) the lubricant almost solidifies, depending on the type of lubricant.

Incubation: The incubation period occurs with the plastic deformation of surface asperities and cyclic contact and stresses accumulate plastic deformation and initiate fatigue cracks.

Nucleation: After fatigue cracks appear, they grow and coalesce. The resulting pit can be up to 10 microns, imperceptible to the human eye.

The particles generated by micropitting are of the order of one micron or smaller, up to 10 or 20 microns. It is usually not possible to eliminate these particles with strainers and they act by polishing the gear surface. Polishing wear is often found when micropitting is present.

Effect of lubricants

The properties of the lubricants, such as base oils, additive chemistry and viscosity all affect micropitting. Micropitting tests show that resistance to micropitting varies from one lubricant to another.

Some lubricants are capable of halting the process once it has started.

Base oils: Lubricants solidify under the high pressure generated in elastohydrodynamic lubrication (EHL) conditions and tractional stress on the surface asperities is limited by the rupture stress of the solidified oil. There are considerable differences between the solidification pressure and the rupture stress of different lubricants and hence differences in their tractional properties.

Polyglycols and esters have molecules with flexible ether linkage and a lower rupture stress value than hydrocarbons. Naphthenic oils are relatively rigid, compact molecules that generate high traction, while paraffinic oils and polyalphaolefins (PAOs) have open, elastic molecules with a low traction coefficient. PAOs and non-conventionally refined oils have less traction coefficient than solvent-refined oils. Many PAOs are blended with esters to increase solubility for additives. Unfortunately, esters are very hygroscopic and so the micropitting resistance of the PAOs may decrease significantly.

Micropitting occurs with both mineral and synthetic oils. At high temperatures, PAO and PAG synthetic oils have thicker EHL films and therefore greater resistance to micropitting than mineral oils with the same viscosity grade and additives. At temperatures ranging from 70°C to 90°C little difference exists between the mineral oils and the PAOs, whereas PAG lubricants have thicker films.

Additives: Anti-scuffing (EP) additives are normally necessary but can be chemically aggressive and may promote the appearance of micropitting. Oils without anti-scuffing additives provide maximum protection against micropitting.

Experiments show widely varying and sometimes conflicting, results regarding the influence of EP additives on the onset of micropitting. Some tests show that oils with additives containing sulphur and phosphorus (S-P) promote the appearance of micropitting while other tests show that these additives increase resistance to micropitting.

The additive activation temperature may be one of the reasons for these conflicting results. If tests are performed at different temperatures, the performance of the additives will be different. Therefore, all micropitting tests should be conducted at a temperature closest to the operating temperature. The generally accepted test is the FVA 54, performed at 90°C and lubricant manufacturers increasingly characterise the oil at 60°C.

According to Dr. Andy Olver, EP additives react with the tooth surface reducing its resistance to surface fatigue and are therefore not a suitable alternative. The additive level should not be lower than 50% of the value of the new oil as it would significantly affect its performance.

Viscosity: Low viscosity oils reduce film thickness which in turn promotes the propagation of cracks. High viscosity oils provide greater resistance to micropitting because of the thicker film and are therefore less likely to promote the propagation of cracks. However, viscosity must be limited since excessively high viscosity can promote oxidation of the oil, loss of energy, more residue, etc. Thus, viscosity should be carefully calculated in order to protect the components from all kinds of problems. Variations in the viscosity above 10% with respect to the value of the new oil should not be permitted.

The influence of additives can ruin the effect of viscosity. Therefore, increasing the viscosity will not eliminate micropitting if the base oil contains aggressive additives.

Particles: Solid particles in the oil that are larger than the EHL film can enter between the gear teeth due to the rolling action. Once they enter, they are subject to high pressure levels. The particles are brittle and break up into smaller particles, some becoming embedded between the gears and others passing through the contacts. Hard particles larger than the film thickness can pass through the contact.

The small particles that enter the contacts cause dents in the gears and promote the formation of micropitting. Particles that have not been removed in the manufacturing process must be eliminated immediately by filters. It is very important that the oil is clean when incorporated into the machine.

Water: Many experiments have shown that oil in the water promotes wear. Hydrogen blistering and embrittlement (a phenomenon caused by hydrogen atoms entering the cracks and fissures of the material, forming hydrogen molecules or combining with a metal) may produce a failure. The maximum water content admissible in gearbox oil must not be over 200 ppm.

Elastohydrodynamic (EHL) lubrication

The film thickness of the oil is determined by the oil's response to the shape, viscosity and velocity of the contact inlet. Higher load causes increased elastic flattening without producing significant changes in the inlet geometry. Therefore, film thickness is insensitive to the load and elastic properties of the material. In contrast, the film thickness is highly influenced by the entrance speed of the oil and by its viscosity. It is also highly dependent on the gear temperature, but not on the higher flash temperature, which usually occurs in the central region. The central region of contact is relatively long. Once inside the central region, the oil can not escape because of its high viscosity, the gap is small and the time of contact very short. Generally, all the oil that enters goes through the contact area as a solid sheet of uniform thickness. On exiting, the oil reverts to its atmospheric properties.

12.7.2   Operating parameters to control micropitting

Load

Load does not have a major influence. High loads do not mean that micropitting will occur in the gears. Dr. Andy Olver reports that the formation of micropitting is influenced by the load.

Velocity

Rolling velocity is very important and beneficial since it increases the entry speed of the lubricant, promoting the formation of films and reducing the influence of the asperity contact. Sliding speed, on the other hand, generates heat and increases the formation of particles.

Temperature

The temperature is critical to film thickness and the activation of the lubricant additives. The equilibrium temperature is established by the balance between the heat generated by friction and the heat dissipated by conduction and convection. The temperature of high-speed gears can be much higher than the temperature of the oil supplied to the gears.

Micropitting resistance decreases with higher gear-tooth temperature. However, the performance of some additives improves with higher temperatures. This is very important when establishing temperatures in laboratory micropitting tests.

Thus, micropitting can be prevented by maximising film thickness, reducing surface roughness (coating the gears, etc.) and optimising the properties of the lubricant by avoiding aggressive EP additives, maintaining the oil clean during its lifetime, using lubricants with a low traction coefficient, etc.

Remember, the first thing to do is to choose the correct type of lubricant and then ensure it is kept clean, dry and at the desired temperature.

12.7.3 Advantages of lubrication

Advantages of automated lubrication

1. Lower costs for repairs, spare parts and lubricant.
2. Improved operating times, less costly downtime.
3. Longer maintenance intervals – one year or longer.
4. Greater bearing life from regular, exact amounts of lubrication.
5. Reduced safety issues associated with hard-to-reach lubrication points.
6. Added corrosion protection from the elements.
7. No wasted lubrication.

12.7.4 Adhesive lubricant for gear teeth

1. Excellent load carrying capacity.
2. Excellent mechanical stability.
3. Excellent duration of the lubricating film.
4. Excellent adhesion even on vertical tooth flanks.
5. Allows extremely long re-lubrication intervals.
6. Prevents seizure and wear on tooth flanks.
7. Outstanding corrosion protection.

12.8 Summary of lubricants

1. Wind Turbines require large amounts of high end lubricants.
2. These applications require regular re-lubrication.
3. Not all lubricants are alike even if they are on the approve list.
4. Efforts are being done to help extend lubrication intervals and lower the amount of labour required.
5. If you are thinking of switching lubricants, make sure you talk to an expert at the OEM or at a lubricant manufacturer.

12.9 High efficiency nanofluid cooling system for wind turbines

Wind turbines, during operation, need to dissipate a large amount of heat, that, if not properly handled, might generate a temperature rise of the electrical and mechanical components and hence a further reduction of the overall efficiency.

High temperatures also contribute to unexpected crash of the generators, which results in very expensive repair costs, particularly, for offshore power plants.

The cooling system of most wind turbines uses a forced flow of external air as heat transfer fluid. Such a flow directly cools the electrical and mechanical components or passes through an air/liquid heat exchanger, usually located on the top/back side of the nacelle. Manufacturers frequently use liquid cooled generators for turbines that operate in harsh environment. These types of generators are more compact than air-cooled ones and characterised by higher electrical efficiency, because of the better cooling and lower drag/friction losses.

The cooling system of most wind turbines usually requires high electric consumption to establish and sustain the airflow and, therefore, it increases the amount of dissipated heat. In addition, in the former case, the airflow can carry a large amount of dust, sand, salt, etc., within the nacelle, whereas in the late case the heat exchanger affects the airflow around the turbine, thus making more difficult its control.

The rise in size of new generation wind turbines increases the heat to be dissipated to make the system work properly and efficiently. This work investigates the potential performance of an innovative cooling system, based on the use of the wind turbine tower as heat exchanger, here after referred to as wind tower heat exchanger (WTHE), coupled with the use of innovative heat transfer fluids, made of a mixture of water and nanoparticles, hereafter referred to as nanofluids.

This solution gives many advantages *vs.* the traditional cooling systems, under the point of view of the contaminations from moisture, salt, sand or other impurities into the nacelle. The new cooling technique can be used both for onshore and offshore wind turbines, but the advantages are more evident in the second case, due to the most severe operating conditions.

The use of nanofluids represents a possible solution to enhance the performance of water-cooled systems. Since 1904, Maxwell proposed to use high conductive particles suspended in a liquid to increase heat conductivity in common heat transfer fluids. Nanofluids are engineered colloidal suspensions of nanoparticles (1–100 nm) in a base heat transfer fluid such as water, organic or metal liquids, etc. Nanoparticles are typically made of chemically stable metals, metal oxides or carbon.

Solid particles have higher thermal conductivity than liquids and, therefore, this contributes to enhance heat transfer, momentum and mass transfer and reduces the sedimentation and erosion. Such enhancement also depends on other factors, such as particles shape, volume fraction and thermal properties. First studies investigated millimeter or micrometer particles sized, but, although revealed some enhancement, their dimensions caused quick sedimentations, abrasions and clogging. Nevertheless, such studies revealed a rise of 20% in thermal conductivity of nanofluids using 4 vol.% of CuO nanoparticles, with average diameter of 35 nm, dispersed in ethylene glycol. A similar behaviour has been observed with Al_2O_3 nanoparticles and better results were obtained by using Cu nanoparticles or carbon nanotubes.

12.9.1 Cooling system configuration

In the proposed cooling system the waste thermal load from electrical generators and mechanical components is dissipated through a WTHE, as Fig. 12.2 schematically shows.

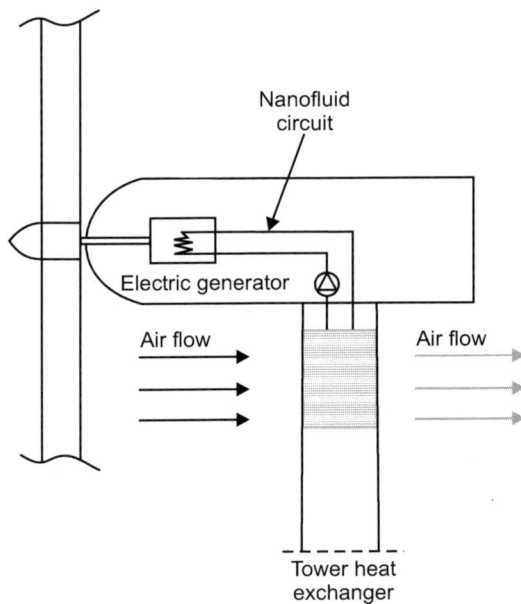

Figure 12.2: Investigated cooling system configuration.

The WTHE is made of a spiral pipe, welded on the internal side of the tower. To reduce the height of tower involved in the heat transfer, each spire has been considered welded side by side one to each other. Besides, to improve heat transfer, a water-based nanofluid with Al_2O_3 nanoparticles are used in the WTHE circuit, instead of pure water. In the present investigation, a 2 MW wind turbine has been studied, whose main characteristics are reported in Table 12.3.

For more details, Fig. 12.3 shows the electric generator efficiency curve and Fig. 12.4 shows the relationship between the generator cooling water flow and the related pressure drop.

Table 12.3: Wind turbine dimensions.

Parameter	Value
Maximum tower diameter	4.15 m
Minimum tower diameter	2.30 m
Tower height	60.00 m
Rotor diameter	76.00 m

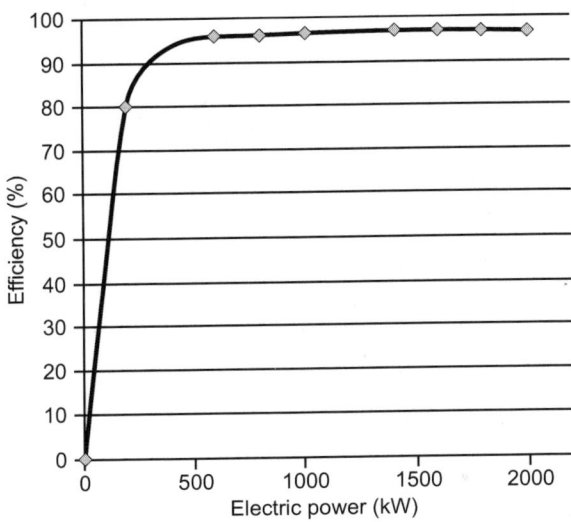

Figure 12.3: Electric efficiency of the 2 MW wind turbine generator.

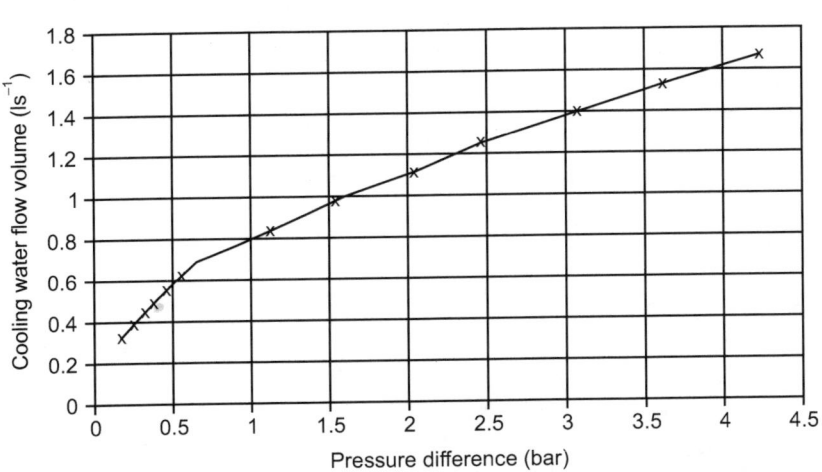

Figure 12.4: Experimental electric generators cooling water flow volume as a function of pressure drop.

12.9.2 Nanofluid characterisation

Nanoparticles characterisation: measurements and models

Mixtures of base heat transfer fluid (water) with Al_2O_3, commercially available nanoparticles, have been tested in order to calculate nanofluids thermal properties for all the investigated operating conditions. The main characteristics of the Al_2O_3 nanoparticles are given below:

1. Spherical shape.
2. Effective density: 3970 kg/m^3.
3. Mean size: 22.91 nm.

In order to predict the thermal conductivity and stability of nanofluid solid-fluid mixtures, many models have been developed, based on different theories, in order to optimise the thermal performance of the system.

Wind energy powering agriculture

13.1   Introduction

Humankind has been using wind energy since ancient times–for sailing, water pumping and grinding. Modern technology, such as a wind turbine is now used also for electricity production in many parts of the world. It's global application has been increasing almost exponentially over the past years.

Wind, the result of global and local temperature difference, represents another source of renewable energy. The governing principle of wind energy is the transformation of wind flow into rotational movements. This is indeed the same principle as for hydropower systems. The power output of a wind energy system is generally estimated by multiplying the available wind speed by the swept area of the rotor. The rotational force can be used either directly (irrigation pumps, etc.), or to drive a generator and produce electricity (Find an animation of a wind pump here). Energy is one of the major parameters for establishing growth and progress of the country, rather the standard of living depends directly upon the per capita energy consumption.

In agricultural systems, energy is available from different sources as human, animal, sun, wind, biomass, coal, fertiliser, seed, agro-chemicals, petroleum products, electricity etc. Energy sources that release available energy directly to the system are classified as direct energy sources. Renewable energy and farming are a winning combination. Wind, solar and biomass energy can be harvested forever, providing farmers with a long-term source of income.

Farmers and ranchers are in a unique position to benefit from the growth in the wind industry. To tap this market, farmers can lease land to wind developers, use the wind to generate power for their farms, or become wind power producers themselves. Farmers and ranchers can generate their own power from the wind .Small wind generators, ranging from 400 watts to 40 kilowatts or more, can meet the needs of an entire farm or can be targeted to specific applications. In Texas and the West, for example, many ranchers use wind generators to pump water for cattle. Electric wind generators are much more efficient and reliable than the old water-pumping fan-bladed windmills. They may also be cheaper than extending power lines and are more convenient and cheaper than diesel generators.

'Net metering' enables farmers to get the most out of their wind turbines. When a turbine produces more power than the farm needs at that moment, the extra power flows back into the electricity system for others to use, turning the electric meter backwards. When the turbine produces less than the farm is using, the meter spins forward, as it normally does. At the end of the month or year, the farmer pays for the net consumption or the electric company pays for the net production. Net metering rules and laws are in place in most states.

Modern agriculture needs modern energy - the two are closely linked. For many developing countries, agriculture is the dominant sector in developing the economy. Increasing productivity and the modernisation of agricultural production systems are the primary drivers of global poverty reduction and energy plays a key role in achieving this. Energy input to modern and sustainable agricultural production and processing systems is a key factor in moving beyond subsistence farming towards food security, added value in rural areas and expansion into new agricultural markets. In many cases, renewable energy technologies and hybrid systems can provide energy services that neatly support the production process, e.g., by providing irrigation (pumps) or post harvest treatment (cooling) or processing (drying, milling, pressing). The requirements of mechanical energy in the agricultural production process are also of critical importance and include human and animal labour as well as fuels for mechanisation, pumping and other activities and indirectly the production of fertilisers and agrochemicals.

Wind can be used to power both mechanical and electric pumps. Mechanical wind-powered pumps use reciprocal non-motorised submersible pumps and require wind speeds of 2.5 m/s minimum up to 4 m/s optimum. Capacity is much lower than for motorised centrifugal pumps, in the range of 1 m^3/hr at depths of 20 metres or more.

Mechanical wind pumps require the availability of local maintenance and repair facilities to be able to respond quickly to mechanical failures. Adequate wind speeds must be present at the location of the wells. One advantage is that they can pump day or night as long as there is sufficient wind and can be used independently of electricity or fuel supplies. A disadvantage is that they must be located directly above the well, a location that may not be optimal in terms of local wind resources. Wind pumps are appropriate in windy areas without other sources of power and only for small irrigable areas.

Wind electric turbines convert the kinetic energy of the wind into rotational mechanical energy that drives a generator to produce electricity, e.g., for pumping water for irrigation. Windmills are positioned for optimal wind conditions, providing greater site flexibility and in addition facilitating electricity production for other uses. Water-pumping applications generally make use of wind turbines with rated output between 1 kWe and 10 kWe. A wide variety of small wind electric turbines is commercially available, with rated outputs ranging from a few tens of watts to 100 kilowatts and is used worldwide to provide electricity in locations where alternatives are unavailable or are too expensive or difficult to provide.

13.2   Hybrid systems

Small-scale hybrid power systems, also a mature technology, are used worldwide. By combining different energy sources (solar-diesel, wind-diesel) hybrids can provide widespread and highly reliable electrical supply. These small hybrid systems are easy to install—no special tools or concrete are required. The term wind hybrid system describes any combination of wind energy with one or more additional sources of electricity generation (e.g., biomass, solar or a generator using fossil fuels). Hybrid system are very often used for stand-alone applications at remote sites.

The combination of renewable energy technologies allows a more balanced electricity supply during day/night and seasonal changes. At most sites wind speed is low, when the sun is shining and reaches higher values on cloudy days. Thus the amount of energy generated by wind energy reaches its maximum in the winter months, while the output of PV-cells is significantly higher in the summer. Other important examples are wind-diesel systems often used in remote areas. A diesel generator will be used as backup, if the electricity demand can not be covered by the installed wind turbines. Regulation and conversion of the available energy sources is a central issue planning a wind hybrid system. Many hybrid systems are used as stand-alone off-grid applications.

13.2.1   Energy storage

Hybrid systems contain an energy storage device to store the surplus energy during times of high energy production, which can be used for supply when production from renewable sources is low (e.g., no wind). For this reason, the size of the device is often described by the period of time in hours h_0 the average load can be covered using the storage as the sole source of energy. Other important characteristics are the overall efficiency of the storage device (determined by the loss of energy during the charge and discharge-process), the output voltage U_b and the maximum permitted discharge. Lead-acid batteries today are the most common technology solution used in hybrid energy systems. There are several alternatives like flywheels, pumped hydro storage, hydraulic storage and fuel cells.

By storing surplus energy and operating as an additional energy source, when production from the RES-source is low, the independence of the hybrid system is increased. If a diesel generator is part of the system, the storage will allow a more efficient management of the given generation units, avoiding emissions by a more efficient utilisation of the diesel generator. Frequent shutdown and restart procedures as well as very unefficient generator loads can be avoided. In this manner, depence on the availability of fuel and thus on fuel price variability is reduced. Regulation by a storage device improves the quality of the supplied power, because variations in the frequency of the current can be minimised and voltage control is available. The degree of this improvement clearly depends on the size of the storage device and the adjustment of the whole system. It must be mentioned, that a storage device significantly raises the initial costs of the hybrid system. Desregarding the type of storage which is chosen (pumped hydro, batteries and flywheels, etc.), the environmental impacts have to be considered. Losses during charging and discharging-processes lower the efficiency of the whole system reducing the positive effect of avoided generator utilisation.

13.2.2   System electronic devices

AC/DC rectifier: In case the energy storage device consists of batteries, the three-phase AC current generated by a wind turbine has to be converted in a DC current for charging. This task is achieved by an AC/DC rectifier of a nominal power x_r corresponding to the rated power of the wind turbine x_0.

DC/DC charge controller: The AC/DC rectifier connects the generating units with the DC/DC charge controller of a rated power of x_c charging the battery system with a charging voltage U_{cc}. Besides the charge controller distributes the incoming energy between the charging process and other DC loads which have to be covered within the hole hybrid system. This description is valid for systems using batteries as energy storage device. For a storage fed by an AC current (e.g., pumped hydro storage) the output of the generating units certainly does not have to be converted. Nevertheless in this case a controlling unit is needed for distribution of energy between storage and system loads.

DC/AC inverter: The energy stored in the batteries has to be reconverted into AC current before it can be used to supply a load. Thus a DC/AC inverter has to be included.

13.2.3   Wind-diesel hybrid systems

The combination of a diesel generator and a wind turbine in a hybrid system a very common and frequently used in remote areas. The following description includes considerations about system design and sizing of the components. Central questions of system designare explained by the example of the wind-diesel hybrid system.

Components

Since an equally distributed power supply by RES-sources is essential for the efficiency of a hybrid system, the wind turbine generator should be applicable for Maximal Power Point Tracking (MPPT). During times of low wind speed the tip speed ratio of the wind turbine must be adjusted by controlling the electromagnetic torque. Thus the wind turbine should be supplied with a modern power electronic control. For wind velocities higher than the rated wind speed of the turbine, pitch-control is an important tool to keep currents and voltages within safety limits. These control mechanisms influence the sizing of the system, because the capacity of controllers and inverters could be decreased including a lower capacity overhead needed to protect the system components against overloads.

The wind generator can either be a syncronous or an induction generator. Often used modern generator types are permanent magnet syncronous generators (PMSG) or doubly-fed induction generators (DFIG). In case of induction generator a source for excitation, either by excitation capacitators or by grid-connection must be available. In modern stand-alone hybrid-systems a direct-driven version of PMSG is the prevailing choice of wind turbine, because it does not require an external DC current for excitation. Problems to keep frequency at 50 Hz during low wind velocities are avoided by modern construction concepts with a large number of poles. The diesel generator should be supplied with a syncronous generator. A first order model with a single time constant can be chosen. The single time constant describes the ratio between fuel consumption and mechanical torque production. The action of the speed governor is controlled by an integral controller gain.

System design and sizing

Load assessment: If the planned hybrid-system is the only source of energy for a village or community, it will be important to categorise the loads which have to be supplied in the village. Medical centres are an example for high-priority loads, while economic and agricultural loads can be labelled as medium-priority loads. Finally domestic supply can be considered as a low priority load in most cases under the assumption, that electricity is a scarce resource. By categorising the prevailing loads and collecting information about the distribution of these loads during the day the project developer creates the base for choosing the size of the system components.

Resource assessment: The distribution of the wind velocities at the proposed sites has to be analysed.

Sizing of the generation units: The essential question to be answered concerns the ability of the combination wind turbine/storage to cover the high-priority loads. Designing a system which covers these loads by RES-sources or storage means the system is generally based on wind energy with the diesel generator as additional or back-up energy source. This operation strategy assures, that a great amount of fuel can be saved, which is important for gaining indepence from high fuel prices and variability. Optimisation of a wind-diesel hybrid system is a complex procedure, because besides the covering of the loads the utilisation of the diesel generator and the storage system must be considered. Diesel generators typically have a minimum load for operation (20–40% of rated capacity) and their efficiency varies significantly with load level. Additionaly frequent on/off-switching of a generator causes severe wear and tear processes and should be avoided for that reason.

The storage systems consist of a battery bank, a bi-directional power electronic converter and a current limiting impedance. It can be considered as a management tool between fluctuating loads, wind energy production and the efficient utilisation of the diesel engine. The charging-discharging rates of the battery technology chosen should be able to cope with these tasks in the system. Commonly used battery types are lead-acid and nickel-cadmium batteries with energy densities of the order of 0.05 kWh/L and 0.1 kWh/L.

Sizing of converters: Converters have a great influence on the overall system efficiency, because they have a certain amount of self-consumption and their efficiency varies with load. At two-thirds of its rated capacity the input-output efficiency of converters is typically about 87–95%, but during times of very low loads (which occur very often especially in domestic supply), the converter efficiency falls rapidly and can reach values under 50%. Thus converter size must be chosen carefully. System failure will be caused, if the converter is chosen to small and overloads occur. If the size of the converter is too big, it will operate on inefficient load levels during long time periods.

Two strategies are implemented to avoid the difficulties of converter-sizing: In certain systems the number of converters is equal to the number of loads assuring that the influence of the converter on the system performance only occurs, if the related load has to be covered. The second strategy creates to separate systems. One DC-System for lighting and other DC-applications and one AC-System, which is only used in case an AC-application has to be supplied.

13.2.4   Costs of wind hybrid system

One main disadvantage of any hybrid system are the high initial cost, caused by the number of necessary parts in the system and especially by the relatively high costs of energy storage systems and – in case of their integration – PV-Panels. In remote areas alternative solutions like grid extensions or dependence on sole diesel generation are expensive as well. While grid extension may result in exorbitant installation costs, pure diesel generation causes a very low initial investment but the maintainance and operation costs are a very substantial factor due to fuel prices and consumption. As a result planning a wind hybrid system requires a comparison of overall costs containing initial installation costs and maintainance and operation costs.

13.2.5 Applications of wind hybrid systems

T/C stations

Stations for telecommunication in remote areas have to be supplied with power during long time periods. Extension of the electricity grid in most cases is a big financial effort, while the supply by sole diesel generators causes additional fuel- and maintainance-costs permanently. Small hybrid systems can be used to reduce fuel-consumption. A small wind turbine may be placed on the relay-mast of the T/C station, avoiding the additional installation costs of a turbine tower. As load variations of a T/C stations are rather low but a steady supply is needed, hybrid systems combining different RES-sources are preferable for this application. A battery storage for system back-up is necessary.

If the T/C station supply should be provided mainly by RES-sources, larger wind turbines on separate towers have to be installed. The inclusion of a PV-System results in reduced variations in RES-output and allows the reduction of the necessary storage size. A well designed hybrid system minimises the fuel costs of the station supply. Several examples have shown the efficiency of fuel savings gained by the application of hybrid systems.

Small desalination systems

Scarcity of potable water is often found in remote island situations, which at the same time often have a good wind potential. In some areas of water scarcity, water desalination contributes a considerable part to potable water supply. Implementing wind-based water desalination units in these areas is a cost-saving and emission-avoiding alternative to running the desalination units by fossil fuels. The techniques described as most efficient for water desalination, are reverse osmosis and mechanical vapour compression. Unfortunately variations and interruptions in power supply are very unfavourable conditions for water desalination plants. For this reason hybrid-systems – based on wind, but allowing a less fluctuating energy supply – combine the advantages of renewable fuel saving energy technologies with the necessary supply security for the efficient operation of small desalination plants. A very promising development of technology are floating systems, consisting of platforms with an installed hybrid system and micro desalination plant, floating near the coast on the water surface and using the preferable offshore wind conditions.

Water pumping

There are many areas with scarce surface water availability but sufficient water resources in appropriate depth for pumping systems. In case of poor infrastructure for water supply, wind turbines can be used as electricity source for these systems. Common aims are providing water for domestic and community supply in remote locations, as well as cattle watering and irrigation. Agriculture still is the economic activity with the greatest water consumption (almost two thirds of worlwide water consumption). The government of India promoted the implementation of wind-pumping system very intensive. The high distribution of small systems has created a well established market for small and multi-bladed wind turbines in recent years.

Besides a shift to Photovoltaic-systems occurred because water scarcity is more often correlated with sun radiance as with appropriate wind conditions. The price-reduction has decreased the prevailing barriers of installation costs of these systems. For the same reasons as mentioned related to desalination plants hybrid-systems have advantages in comparison to the sole use of wind or PV. The greater availability of hybrid systems is an important characteristic, because a reliable water supply not at least economically essential for the growth of plants, watering of cattle and not at least the sufficient supply of the inhabitants of remote areas.

13.2.6 Choosing the right energy source

Different pump technologies are often flexible regarding the type of energy source powering them. There follows an overview of the different energy sources available for motorised pumping and the definition and characteristics of each.

Grid electricity

Where grid connection is available an electric pump can be powered directly. However, the cost, availability, reliability and quality of local electricity supplies determine the use of this energy source.

Fuels for combustion engines

Motorised pumps can be powered by fossil fuels (diesel, gasoline), either through generators that create electricity, or by transmitting power to the pump through a drive belt and vertical rotating shaft. In addition, some submersible pumps (i.e., progressing cavity pumps) operate by direct displacement, like piston pumps. The pumps tend to be more expensive but nevertheless more efficient than centrifugal pumps.

13.3 Wind-powered water pumping systems for livestock watering

Water supplies such as wells and dugouts can often be developed on the open range. However, the availability of power supplies on the open range is often limited, so some alternate form of energy is required to convey water from the source to a point of consumption. Wind energy is an abundant source of renewable energy that can be exploited for pumping water in remote locations and windmills are one of the oldest methods of harnessing the energy of the wind to pump water.

Windmills generally consist of two types, with the classification depending on the orientation of the axis of rotation of the rotor. Vertical-axis wind turbines are efficient and can obtain power from wind blowing in any direction, whereas horizontal axis devices must be oriented facing the wind to extract power. Most windmills for water-pumping applications are of the horizontal-axis variety and have multi-bladed rotors that can supply the high torque required to initiate operation of a mechanical pump. Windmills can also be used to generate electricity, but electricity-generating units usually consist of vertical-axis rotors or highspeed propellor rotors, due to the requirement for low starting torques. Figure 13.1 shows a typical water-pumping windmill.

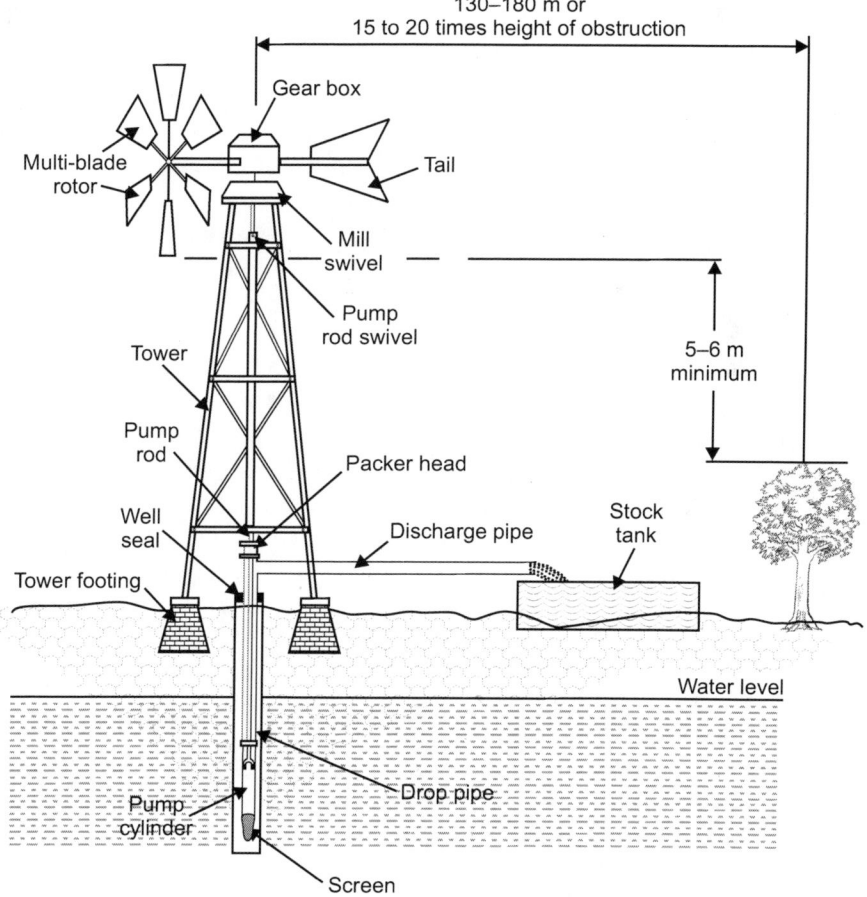

Figure 13.1: Typical windmill water pump.

13.3.1 Choosing a location for a windmill

The primary consideration in choosing a site for a windmill is whether there is sufficient wind for such a device to be feasible. Although wind speed is monitored at a number of locations throughout Canada, this information is of limited usefulness because vegetation and other geographic characteristics can result in large variations in available wind power over short distances. Obtaining site-specific measurements of wind speed and duration during the period over which water pumping is required is the only reliable way of determining whether a wind-powered pumping unit will be a viable option. To take such measurements, an *anemometer* is required. Economical hand-held anemometers are available, but their use requires that a considerable amount of time be spent on site to establish meaningful records. A better way of gathering wind data would be to mount an anemometer (with an automated data recording device) on a tower similar in height to the proposed windmill for the entire period of interest.

Typical television or radio antenna towers can be used and some are available as portable, trailermounted units. It may be possible to rent portable towers in some areas. Windmills used to generate electricity to power an electrically-powered pump can be located away from the pumping unit and windmills that power an air compressor which operates an air-lift pump can also be located away from the pump. However, most windmills are designed to operate a reciprocating piston-type pump and must be located directly over the water source (usually a well).

To ensure that the windmill receives a free flow of air from all directions, the rotor of a windmill should be located at least 5 to 6 m (15 to 20 feet) higher than any obstruction within about 130 to 180 m (450 to 600 ft.) of the windmill site. In fact, wind speed generally increases with altitude, so the tower should be as high as reasonably possible, regardless of the presence of obstructions. Topographic effects, such as confined draws and hills, should also be considered.

How much water can a wind-powered pump deliver?

The amount of water a wind-powered water pumping system can deliver depends on the speed and duration of the wind, the size and efficiency of the rotor, the efficiency of the pump being used and how far the water has to be lifted. The power delivered by a windmill can be determined from the following equation:

$$P = 0.0109 \, D^2 V^3 \eta$$

where, P is power in watts, D is the rotor diameter in metres, V is the wind speed in kilometres per hour and η is the efficiency of the wind turbine. As can be seen from this expression, relatively large increases in power result from comparatively small increases in the size of the rotor and the available wind speed, doubling the size of the rotor will result in a four-fold increase in power, while doubling the wind speed will result in an eight-fold increase in power. However, the efficiency of wind turbines decreases significantly in both low and high winds, so the result is that most commercially-available windmills operate best in a range of wind-speeds between about 15 km/hr and 50 km/hr.

13.3.2 Types of pumps used with wind mills

If the windmill is used to generate electricity to power an electrically-powered pump, it will probably be necessary to store the electricity in batteries due to the variability in generation. Therefore, a pump powered by an electrical motor for use in conjunction with a windmill that generates electricity should have a Direct Current (DC) motor. For such systems, it is important to use good-quality deep-cycle batteries and to incorporate electrical controls such as blocking diodes and charge regulators to protect the batteries.

The most common type of pump used with windmills is the positive-displacement cylinder pump driven by a reciprocating rod connected to a gear box at the windmill rotor. The performance of these pumps can be enhanced through the addition of springs, cams and counterweights that alter the stroke cycle and off-set the weight of the drive rod, thereby reducing the starting torque and allowing the system to perform better in light winds.

An alternative to the traditional cylinder pump is the air-lift pump. The air-lift pump is a type of deep-well pump, sometimes used to remove water from mines. It can also be used to pump a slurry of sand and water or other 'gritty' solutions. In its most basic form this pump has no moving parts, other than an air compressor driven by the windmill and the efficiency of the air compressor is a prime factor in determining the overall efficiency of the pump. Compressed air is piped down the well to a foot piece attached to the discharge pipe. As air is discharged into the water column in the discharge pipe, a two-phase mixture of air and water is formed that is less dense than the surrounding water in the well. This apparent density difference is what causes water to rise in the discharge pipe.

Air-lift pumps can lift water at rates between 20 to 2000 gallons per minute, up to about 750 feet. The discharge pipe must be placed deep into the water, from 70% of the height of the pipe above the water level (for lifts to 20 feet) down to 40% for higher lifts. This is the most significant drawback to air-lift pumps, because many wells do not have the required depth of standing water. An advantage to this kind of pump is that the windmill can be located away from the well and the windmill/air-compressor combination can also be used to aerate dugouts.

Other considerations

The availability of wind energy is highly variable and is likely to be intermittent. As a rule-of-thumb, expect an average of 6 to 8 hr per day of water pumping at a rate specified for a 25 km/hr (15 mph) wind. Hand pumping can be done on some windmill pumps in emergencies, but wind-powered pumping units should be used in conjunction with storage facilities capable of meeting three or four days water demand as a back-up supply during periods of low wind.

The cost of a wind-powered pumping system will vary according to its capabilities. One of the attractions to wind-powered pumping systems is their simplicity and robustness. It is prudent to check all rubber diaphragms annually, and, on compressed-air models, to check all valves. The anchoring system for the windmill tower should also be checked to ensure that the windmill is not toppled during high winds. As with any pumping system, a wind-powered pumping system should be checked regularly to ensure that cattle have an adequate supply of water.

Wind-powered water pumping systems are just one of many options available to producers interested in managing their rangelands, providing improved water quality for their livestock and protecting their water supplies.

13.4   Calculating the wind energy

To calculate the energy yield of a wind turbine, multiply the generated power by the duration during which the wind turbine is running. Since wind turbine produces different power outputs at different wind speeds, the mentioned multiplication is a little more complex. We can therefore look in detail at the wind conditions in a specific location. Such data can be obtained from wind/weather measurement stations, or from wind atlas if available for the site of interest.

Figure 13.2 shows an estimation of the annual yield of wind power in which wind speeds are predominant at the location. This graph is called the frequency distribution. One can see that a wind speed of about 5 m/s occurs more than 1000 hr a year, while stronger wind speeds are less frequent throughout the year.

The second input for calculating the annual energy yield of a wind turbine is the power curve of the wind turbine. The power curve shown in Fig. 13.3 is specific for every turbine model and normally provided by the manufacturer. It says at which power output the turbine generates which wind speed. When multiplying the hours and the corresponding power output for every wind class (1m/s, 2m/s, 3m/s) the result shows the energy production per wind class. To get the Annual Energy Production (AEP) of the turbine, sum up the

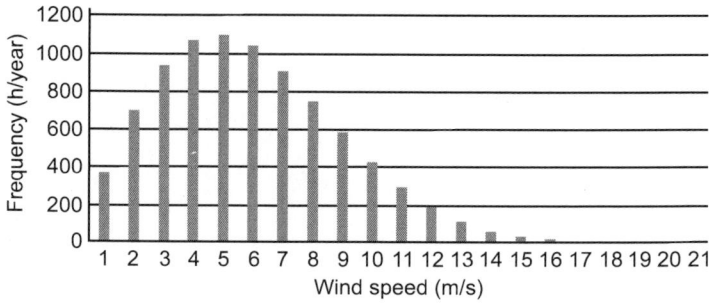

Figure 13.2: Estimation of the annual yield of wind power.

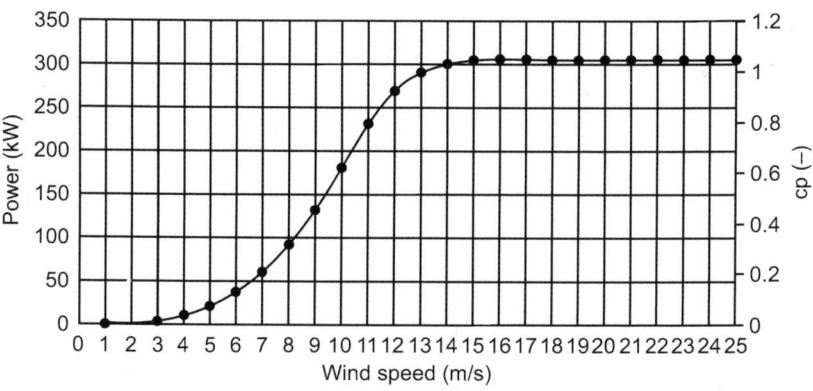

Figure 13.3: Power curve of the wind turbine

energy yield (Ei) from every wind speed. These calculations can be done with software support (e.g., Excel) or manually.

To calculate the power output of a wind turbine, check out the following equation:

$$P = 0.5 \rho A v^3 Cp \eta$$

P = Power [W]

A = Swept area of blades [m²]

v = Wind speed at hub height [m/s]

ρ = Density of air (1.29[kg/m³])

Cp = Power coefficient (Betz's coefficient) (59%)

η = System efficiency

Wind energy based desalination processes and plants

14.1   Introduction

Water is a valuable natural resource and its shortage is a serious problem being faced by many areas of the planet. Decision making on the water supply method includes technical and economic evaluation of various alternatives, taking into account the urgent character of the problem and the need for its sustainable solution. Desalination of brackish and sea water has become one of the most widely applicable methods to meet water demand and it is today widely applied in areas with limited water resources. One of the most promising desalination methods is based on Reverse Osmosis (RO) phenomenon. A critical issue in water desalination is the high energy demand and, more specifically, electricity for RO desalination units. RO is the desalination process with quite low energy requirements.

14.2   RES based desalination processes

The use of Renewable Energy Sources (RES) for the operation of desalination plants is a feasible and environmentally compatible solution in areas with significant RES potential. The main driving forces for applying RES in desalination plants are the seasonal variability in water (and energy) demand, usually occurring when renewable energy availability is high, the limited availability of conventional energy supply in remote areas, the sufficiency of RES in islands, the technological advancements being achieved in desalination systems, the limitation of environmental impacts of conventional desalination systems and the relative easiness of the plant's operation and maintenance compared to conventional energy ones.

Especially for the Greek islands, where there is water and energy shortage and, at the same time a high RES potential, the RES based desalination plants comprise a promising technically feasible and financially attractive solution.

To that end, a lot of research and development work has been carried out and the problem of the optimal configuration/combination of a RES energy source with a desalination plant attracts the interest of many researchers and construction and engineering companies.

The best coupling of RES to desalination systems is determined from various criteria, such as the system's efficiency, the investment and operational cost, availability of operational personnel, the suitability of the system to the characteristics of the location, the possibility for future increase of the system capacity, etc.). The selection of the appropriate RES desalination technology depends on a number of factors, including:

1. Required quantity of potable water (plant capacity).
2. Feed water salinity.
3. Remoteness.
4. Availability of grid electricity.
5. Technical infrastructure.
6. Type and potential of the local renewable energy resource.

Various combinations of RES and desalination systems have been proposed and implemented, each one with its own characteristics and suitability under certain criteria. Desalination systems driven by wind power are the most frequent renewable energy desalination plants. Coastal areas have a high availability of wind power resources and wind power is the most competitive renewable energy technology in power generation. Therefore, wind powered desalination is a promising alternative. The idea to use wind power as an energy source for desalination is not new. Wind conditions, for example, in coastal areas are often in favour of this desalination system.

More specifically, wind energy can be used efficiently on condition that the average wind velocity is above 5 m/s. This makes wind powered desalination a particularly interesting option for windy islands, both for the solution of their energy supply problem and for the operation of sea water desalination plants. The new generation of small and medium sized wind turbines that has been developed in the past years offers a high amount of reliability in service combined with low investment costs.

Two types of applications are usually referred as wind energy and desalination couplings. The first type concerns the coupling of the wind generator and the desalination plant on a small size autonomous electricity grid. The second concerns the direct coupling of these two for the sole purpose of production of water. Desalination plants using membrane technologies are available in a wide range of capacities. As far as the recommended RES – desalination combinations are concerned, it is considered that wind desalination is most suitable for small (1–50 m³/day) and medium (50–250 m³/day) scale plants.

14.3   Energy issues in desalination plants

All desalination systems use energy and, in fact, the energy consumption is one of the most important elements in determining water costs. About 0.7 kWh/m³ is theoretically the minimum energy required to obtain fresh water from seawater. For RO systems the energy consumption is in the range of 5–10 kWh/m³ without energy recovery (large production plants) and 3–4 kWh/m^3 with energy recovery. Recent developments in wind turbine technology mean that wind power can now be regarded as a reliable and cost-effective power source for many areas of the world.

Wind turbines may be classified depending on their nominal power 'N_o' as very small (N_o<10 kW), small (N_o<100 kW), medium sized (N_o<0.5 MW) and large (N_o>0.5 MW) ones. All are based on mature technologies and they are commercially available except for the very large power systems, which still require several adjustments.

The basic assumptions for the required calculations concerning the energy efficiency of the wind turbines with or without an energy storage system may be considered as following:

For a wind turbine with a nominal power of N_o kW, we expect an energy production 'E' in the order of magnitude of '$E = CF.N_o.8760$' kWh/year. Note that the installation capacity factor 'CF' usually varies between 20% and 30%. Depending on the type of desalination plant, the required amount of energy per m^3 of potable water will also be given. Therefore, we may have a series of alternatives concerning the installed power of the wind turbine and the combined capacity of the desalination plant.

Many other parameters should be taken into account in this design issues, such as the possible losses of an energy storage system or the availability of a water storage system. The variable nature of wind power is not a problem as for as water availability is concerned, because water can be stored inexpensively even for long periods of time without deterioration. With a plant that is dimensioned according to the local wind conditions, water becomes available any time. However, variable wind power may cause operational problems in the system's operation and this is one of the most serious issues to be resolved in the design and implementation of this type of projects.

14.4 Environmental impacts of RES based desalination plants

Desalination plants cover the needs of remote areas in water. Usually they are implemented as a result of an analysis and alternative solutions evaluation amongst various possible solutions for water supply. For example, on several Greek islands, fresh water requirements are covered by the construction of large dams or ground reservoirs of desalination plants. In smaller islands, the only available solution is the transport of fresh water by ship, with high costs and improper hygienic conditions. All these water supply methods cause a spectrum of environmental impacts, more or less serious depending on the type of the project, its location and scale.

The main environmental impacts of an RO desalination plant are the following:

1. Noise disturbance.
2. Optical disturbance.
3. Land use.
4. Interference with public access in the coast.
5. Abstraction of brackish groundwater.
6. Discharge of brine - a concentrated salt solution that may be hot and may contain various chemicals on coastal or marine eco-systems or, in the case of inland brackish water desalination, on rivers and aquifers.
7. The emission of greenhouse gases in the production of electricity and steam needed to power the desalination plants in case the energy provided is from the grid and fossil fuels are used to generate it.

The main positive environmental impact from desalination is that it reduces the pressure on conventional water resources. In particular, seawater desalination can help to relieve the pressure on overexploited coastal aquifers. In case that it is decided also to built a RES unit, either to cover only desalination unit's needs or to cover also energy demand of the area, then a strategic environmental assessment is needed that will evaluate in a integrated manner the impacts of all major projects being included.

14.5   Wind based desalination − Operational issues

One of the problems of utilising wind power in process applications is the variable nature of the resource. While the wind is relatively predictable it is seldom constant and there will be periods when there will be none at all. The storage of wind energy in the form of electrical power is really only practical when small amounts are involved. Storage batteries increase the total investment cost therefore, to run a process of any magnitude on stored electrical energy is not a practical proposition.

However if the product of the process can be stored inexpensively then it may be practical to oversize the process equipment to allow for downtime. Water can be stored for long periods of time without deterioration and the storage vessels are relatively cheap.

Variable power input force the desalination plant to operate in non-optimal conditions and may cause operational problems. To avoid the fluctuations inherent in renewable energies, different energy storage systems may be used.

The only matters that would require some careful design would be the relative sizes of the wind turbine and the RO plant and the cutin and cut-out criteria for the RO plant to avoid excessive startup and shutdown cycles.

The intermediate energy storage system would be necessary, but it would reduce the available energy and would increase the cost of the plant.

The main drawback of RO in remote areas is the complex pre-treatment, the requirement of skilled workers, chemicals and membrane replacement.

For the operation of a wind-powered desalination plant, it is most important to have a plant that is insensitive to repeated start-up and shutdown cycles caused by sometimes rapidly changing wind conditions. Reverse osmosis is, with regard to pre-treatment, membrane fouling, after-treatment and efficiency of the high pressure pumps, a process that is rather sensitive to a stop and start operation. In order to stand the discontinuous mode of operation, a new reverse osmosis membrane can be developed, incorporating advantages of both spiral wound and plate and frame designs.

14.6   Wind based desalination – Design issues

The main design variables that affect the design of an wind - RO system are:

1. The water demand and, therefore, the RO plant's capacity.
2. The location that the wind turbine and the desalination plant will be installed (required sitting, altitude etc.).
3. The feed water salinity.
4. The wind speed distribution.
5. The configuration of the energy system.
6. The water storage capacity.
7. The available power distribution.
8. Desalination unit energy consumption.
9. The salt rejection.
10. The operating pressure.
11. The permeate flux, both in terms of overall product rate and specific rate (per unit membrane area).

The various alternative wind - RO configuration possibilities are the following:

14.6.1   Systems with back-up (diesel/grid)

In these systems, an additional energy source is provided (a diesel-powered generator or even the local grid) so that the power supplied to the RO is constant. The back-up generation complements the power generated from the wind turbine to match the RO unit power consumption. The main benefit of these systems, as in any hybrid wind-diesel configuration is the achievement of fuel savings, which may increase the generator availability and reduce overall energy costs.

Systems without back up

Systems without an external energy source can be divided into two categories, with emphasis on the RO unit operation systems which run under approximately constant operating conditions and those that experience variable operational conditions.

Near constant operating conditions

In this case an attempt is made to operate the RO unit with approximately constant operating conditions.

Use of an energy storage device

Energy storage devices are employed to accumulate energy surplus during periods when the power generated by the wind turbine is greater than the load demand from the desalination unit. This surplus would then be used later when the generated power is insufficient to meet the load demand. One common way of storing the surplus energy is by using batteries or water pumping systems. Storage sizing should be considered in the design stage. In addition, capital and maintenance costs should carefully be assessed.

14.7   Cost analysis of wind based desalination

The most promising potential market for wind powered RO is in present or potential future island tourist developments in places such as the Mediterranean islands, the Pacific Islands, etc. Generally, if wind powered electricity generation is an economic proposition in any of these places and water is scarce (which it usually is), then wind powered reverse osmosis should also be economical. It is unlikely that energy storage would prove economical in these larger systems, although energy recovery for seawater plants would almost certainly be so.

In general, the cost reduction of renewable energy systems has been significant during the last decades. Therefore, future reductions as well as the rise of fossil fuel prices could make possible the competitiveness of seawater desalination driven by renewable energies. Product yield depending on wind power available are shown in Fig. 14.1

For a given wind farm installed capacity (with a particular type of wind-turbine) and a given wind regime, there exists, from an economics point of view, an optimum nominal production capacity for each plant, that needs to be specified in each case under consideration. In this context, a wind farm with a nominal power of 460 kW and a wind regime (in the area of Pozo Izquierdo, proposed for its installation in Gran Canaria) with an annual average speed of 7.9 m/sec 10 m above ground level, would give rise to an optimum number of RO plants of 11, each with a capacity of 100 m³/d. However, for

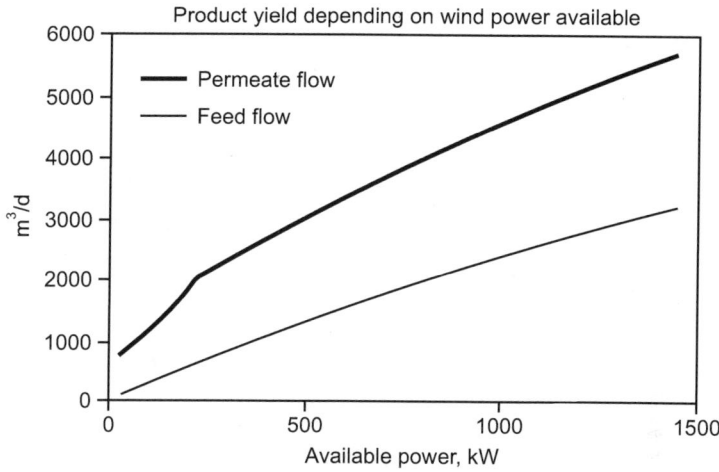

Figure 14.1: Product yield depending on wind power available.

technical and economical reasons the decision can be made to use eight RO plants, each with a capacity of 25 m³/d. Water cost components for a wind powered RO plant are shown in Fig. 14.2.

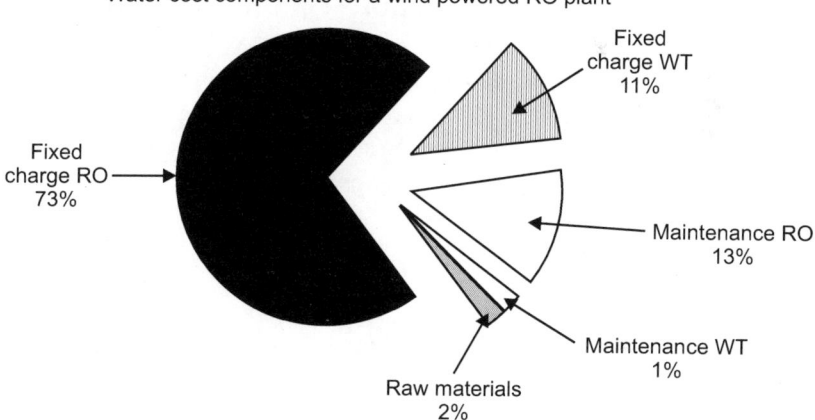

Figure 14.2: Water cost components for a wind powered RO plant.

The average installed costs of seawater RO plants are in the range of $1000 to $1500 (as on 2013) per cubic meter per day capacity.

The economics of a combination of a wind turbine with an RO plant is helped by the fact that water is a storable commodity.

Factors affecting water production cost in wind based desalination plants are shown in Table 14.1.

Table 14.1: Cost items of a wind based desalination plant.

Direct capital cost
Cost of land
Cost of wind turbine
Cost of energy storage systems
Cost of the RO plant components

Annual operating cost
Electricity cost
Manpower cost
Maintenance and spares cost
Chemicals cost
Membranes replacement

The water cost of a wind brackish water reverse osmosis unit (large system, about 250 m^3/day) is of the order of 2 Euro/m^3 (as on 2013). The implemented in Tenerife, Spain included for a 200 kW wind turbine, which would operate on average wind velocity 7.5 m/s, with an expected yearly energy yield around 600 MWh. This amount of energy is capable of producing over 200 m^3/d water.

14.8   Wind based desalination – Implementation issues

The practical experience on wind powered RO systems has been with relatively small capacity systems. There have been a number of attempts to combine wind energy with RO. A number of plants have actually been operated. However, most of them were of small size, mainly for research purposes. Therefore not many conclusions have been reached in terms of expertise and know how. It is still difficult to control the usage of wind in a cost effective way. Coupling of a variable energy supply system, as mentioned earlier, to a desalination unit requires either power or demand management and there is not much experience on it. However, the prospects of this combination are high mainly due to the low cost of wind energy. The operational experience from early demonstration units is expected to contribute to improved designs and a large number commercial systems are expected to be implemented.

14.9   Examples of wind based desalination plants

In this section we will describe two examples of wind based desalination units installed and operating in the Greek islands.

14.9.1   Milos desalination plant

In a Greek island called Milos and belonging in the Cyclades complex, a wind based desalination unit has recently been installed and operates since summer

2007. The unit has a capacity of 3000 m³/day. At the moment it operates in a daily production of 2000 m³ of potable water. This is a private investment that has been subsidised by the state. The water is sold to the municipality of Milos, in a continuous effort to solve the urgent water shortage problem, especially during the summer months. The contract that has been signed between the private company and Milos. Municipality refers to a selling price of the water almost 1.8 euros/m³.

The entire plant includes:

1. The desalination plant.
2. A wind turbine of 600 kW.
3. The storage tanks (capacity 3000 m³).
4. The remote control system.

Before the installation of the unit, water was transported from Athens at a very high cost and very poor quality. The implementation of this novel project has improved the quality of life of the island in many respects.

The sitting of the unit in a very touristic island as Milos could be a major problem, mainly because of the optical and noise disturbance. Therefore, the unit has been located on a hill that is not apparent from the most island villages.

14.9.2   Floating wind turbine/desalination plant

The first floating wind turbine/desalination plant in the world has been developed by a number of scientists/engineers with an academic and professional origin and lead by the University of the Aegean. Two of the most pressing environmental challenges of today – energy production and water supply – have been met with an innovative and practical solution to meet the water needs of Greek islands.

The Floating Autonomous Environmental Friendly and Efficient Desalination Unit (FAEFEDU) is designed to produce potable water from sea water and do so by generating its own power through wind turbines on board. The unit sits on a special floating 20 × 20 m platform with a height of 8 m for a cylinder and a 22 m tower and can adapt to any weather conditions. Water production is more than 70 cubic meters per day—enough for the needs of about 300 people.

In order to achieve the largest possible energy and desalination production, scientists focused on minimising the scale and polluting effects on the central unit, increasing the energy efficiency of the cycle. In addition, because the unit is autonomous, it is not required to be connected to the national electrical grid. Since the unit is portable, it can be stationed away from populated centres and be placed wherever needed, on a seasonal basis for instance, to service the needs of islands that have an enlarged population during summer months. In addition, the unit can be repositioned to take advantage of changing weather conditions. FAEFEDU is fully autonomous, has an advanced automatic control system, operates unmanned and can be tele-operated and monitored remotely. The innovative system also eliminates any destructive land-based environmental interventions, since no roads or land construction are needed and no waste is produced. The Greek research community was heavily involved in the project, coordinating various disciplines to fulfil the diverse needs of the unit. In addition, the Greek shipbuilding sector contributed with its vast know-how to produce the project at an attractive cost. Equipment providers from Germany, Sweden and other European countries also participated. The unit, the first of its kind in the world, was included in the 'Natural Environment and Sustainable Development' section of the Operational Competitiveness Programme of the Greek state and was co-financed by the European Fund for Regional Development and domestic Hellenic sources.

To sum up, the reverse osmosis technique is the most suitable for use in stand-alone wind- powered desalination systems. These systems are very valuable for regions like the Mediterranean islands, usually facing scarcity of potable water and lack of conventional energy sources, but do have at their disposal exploitable wind energy resources.

The financial performance of wind-powered desalination are also favourable. The costs are similar with what is expected for a conventional desalination system, proving to be particularly cost-competitive in areas with good wind resources that have high costs of energy. It can be concluded that wind-powered desalination can be competitive with other desalination systems, providing safe and clean drinking water efficiently in an environmentally responsible manner. Now that significantly larger and more reliable wind turbines have become available, wind powered desalination is poised to make the breakthrough into commercial applications. The actual ratio of wind turbine size to RO plant might be used in any instance should result from an optimisation making use of data that will be site specific for both the wind turbine and for the RO plant.

Role of nanotechnology in non-conventional energy technology

15.1   Introduction

Nanotechnology has opened the floodgate of future opportunities in enhancement of efficiency of non-conventional energy producing technologies. The nanotechnology has touched every facet of entire spectrum of solutions in non-conventional energy. The major impact of nanotechnology is in the use of nano materials which range from nano polymer to metal composites these materials can be tailor made to suit the requirements for particular applications such as solar cells and other direct energy converting technologies. The nanopolymers have revolutionised the construction of wind mill turbines and rotor blades. The low per unit weight and high stress taking capability has been instrumental in creating high efficiency and low energy wasting wind electrical energy generating equipments.

The thermal energy conversion ratio gets geometrically amplified in case of solar photovoltaic cell systems as it decisively intensifies the electron release and photon transformation. The chapter discusses the options, materials and technologies in direct solar application, indirect heat conversion system, wind energy conversion module modification. It also touches the future of hydro power energy production and its control systems and their modification due to impact of emergence of nearly superconducting metal composites in nano technology field.

Nanotechnology unbolts many doors which embrace pressing problems associated with social and environmental issues. It provides alternative solutions to many of the problems faced today by us. It assists by manipulating data at the atomic and molecular level and supramolecular level. Breakthrough in nanotechnology unfasten the probability of strolling beyond our current alternative sources for energy supplies by broaching technologies that are more efficient, inexpensive and environmentally sound.

In the original sense nanotechnology refers to the projected ability to construct items from the microscopic to macroscopic level in coherence with the increase in the efficiency. India may contribute to 25% towards the advancement of nanotechnology.

Nanotechnology boosts the solar efficiency by approximately 80% hence escalating the energy generation, transformation and distribution.

15.2 Salient aspects of nanotechnology

1. Nanomaterials possess unique, beneficial chemical, physical and mechanical properties, they can be used for a wide variety of applications.
2. These applications include, automobiles, ceramics, weapons, weapon platforms, medical sector, protective films, hard cutting tools, airframes.
3. Nanocoated wear resistant drill probes for example can be utilised for the optimisation of lifespan and efficiency of the systems for the development of oil and natural gas deposits or geothermal energy and thus saving the cost.
4. Nanotechnology could contribute to the optimisation of the layer design and the morphology of the organic semiconductor mixtures in the component structures.
5. High duty nano materials are very much suitable for lighter and more rugged rotor blades of wind and tidal power plants as well as wear and corrosion protection layers for mechanically stressed components.

15.3 Types of nanomaterials

The nano materials are used in various shapes and sizes and they are produced tailor made to suit the requirement. The main type of materials are:

1. Carbon black.
2. Silica fumes.
3. Clay.
4. Metal/alloys/composite.
5. Ceramics.
6. Polymer composites.

The above materials are available in the following forms:

Nano ceramic powders: These are solid powders which constitute the most important segment of the whole nano structured materials. The powders constitute more than 50% of the total nano structured materials.

Nano tubes: These are the single or multilayered tubes of conductors and semiconductors. They are strong materials and with high thermal and electrical conductivity.

Nanocomposites: Generally they are polymer based with nano sized fillers used in various applications such as bearing, gear boxes, etc.

15.4 Applications of nanotechnology in energy sector

Nanomaterials assist in higher volume of energy transmission without significantly adding to the losses and wearing away of the parts of the associated with the energy transmission/distribution system. It also exacerbates the carbon dioxide emissions and hence help provide a cleaner energy production and distribution. The transfiguration of primary energy sources into electrical energy, heat and kinetic energy requires utmost efficiency. Efficiency increases specially in the case of fossil fuels and steam power plants. However the escalated efficiency compels higher operating temperatures and hence heat resistant turbine blade materials when considering the probe of wind energy turbines.

However refinement is plausible if the nano-scale heat and corrosion protection layers are employed for turbine blades in power plants or aircraft engines to enhance the efficiency through elevated operating temperatures or the application of light materials. Nano structured semiconductor materials with optimised boundary layer design contribute to increased efficiency that could pave the way for broad application in the utilisation of waste heat for example in automobiles and human body diffused heat.

The utilisation of nanotechnologies for the enhancement of electrical energy stores like batteries and super capacitors proves to be of the most beneficial use. Due to the high cell voltage and the outstanding power energy density the lithium ion technology is regarded as the most promising venture.

15.4.1   Nanotechnology in the sector of solar energy

Conventional solar cells based on photovoltaic technology have come a long way in recent years, but they're still missing a big chunk of the electromagnetic spectrum. The silicon semiconductors in a solar cell are geared toward taking infrared light and converting it directly to electricity. Meanwhile, the visible spectrum is lost as heat and longer wavelengths pass through unexploited. A new nano-material being developed by a group of researchers spread across the country could act as a 'thermal emitter,' making solar power significantly more efficient by withdrawing more of that wasted energy. And hence it increases the utility margin of the exhausted heat by a exorbitant exponential curve. The infrared part of light is relatively easy for conventional high-efficiency solar cells to convert to electricity and the thermal emitter approach works within that framework. A thermal emitter isn't a parallel system for deriving electricity directly from the sun's rays. Instead, this is an application or so-called thermo photo voltaic principals.

Thermo photovoltaic in this context refers to the production of heat from light. The thermal emitter consists of two parts, the first being a tungsten-based absorber that heats up when exposed to light. The emitter component takes that heat energy and uses it to output infrared light, which silicon semiconductor solar cells are already able to absorb. By doing something with all the other wavelengths of light cascading toward solar panels, researchers have estimated a theoretical 80% efficiency rating—much higher than the mid-30s where most silicon-based solar panels are stuck. However, past experiments with thermal emitters found that the increase in performance in the real world was a mere 8%, which hardly justifies the increased cost and maintenance. This turns out to be due to the extreme temperatures involved. The delicate 3D structure of thermal emitters have traditionally not held up to temperatures above 1800°F (about 1000°C). When the entire purpose of a material is to get hot, you want it to thrive at high temperatures, not fall apart.

To address this issue a new method has been employed under which tungsten is used as a thermal emitter. The tungsten structures were coated with a nano layer of hafnium dioxide, a ceramic which added significant structural durability at high temperatures. Whereas raw tungsten absorption surfaces would break down at 1800°F, these nano-coated surfaces operate without issue for 12 hr at that temperature. At 2500°F the material still lasts an hour before breaking down. This approach to improving solar cells is appealing for a variety of reasons. Both tungsten and hafnium dioxide are extremely plentiful and safe to work with. Thermal emitters also work with existing solar cell technology, making it simple to add them to existing systems.

15.5   Nanomaterials in wind power

Out of all the renewable sources of energy the hydropower is the most utilised source and wind power is the next most tapped resource. Conventionally a wind turbine constitutes of the following parts:

1. High mast/tower.
2. Turbine blades.
3. Electrical energy generator.
4. Power system for stepping up the voltage up to transmission level.

The nanotechnology has already created a sizeable footprint in the field of wind energy generation through the usage of advance nanomaterials. The various nanomaterials which are employed as an alternative to the conventional materials are listed in Table 15.1 and can be compared to their counterparts:

Table 15.1: Nanomaterials employed as alternative to conventional materials.

Parts of the system	Conventional material	Nanomaterial	Advantages over the preceding
High mast/tower	Steel shell	Metal alloy composite	(a) Light weight (b) High strength
Turbine blade	Aluminium/Tungsten alloy	Nanocomposite high strength carbon fibre	(a) Extremely light weight

(Cont'd...)

Parts of the system	Conventional material	Nanomaterial	Advantages over the preceding
			(b) Weather proof (c) Easy modification of aerodynamic profile
Electrical energy generator	Copper/Aluminium windings on silicon core	(a) Nanomagnetic materials (b) Nano multilayer rods	(a) No eddy current losses (b) Minimum generator losses (c) High efficiency
Power systems	Copper/Aluminium semiconductors, conductors	Nano-magnets	(a) High efficiency (b) Heat resistant (c) Precision control

Figure 15.1 shows wind energy power systems employing nanomaterials.

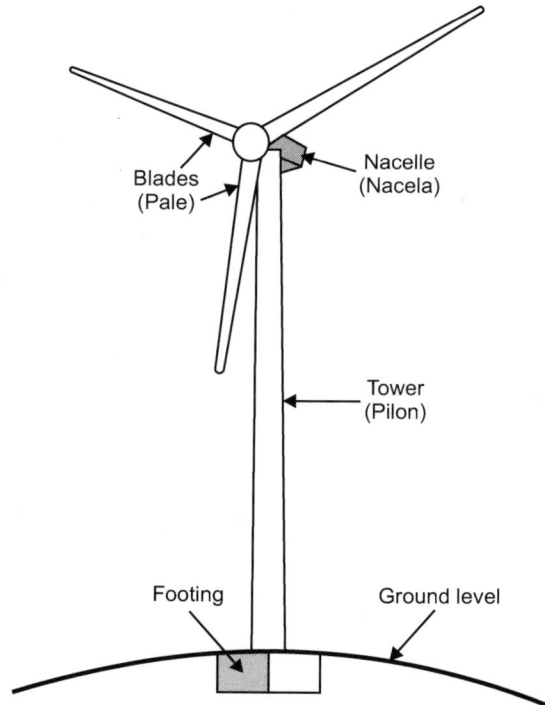

Figure 15.1: Wind energy power systems employing nanomaterials.

Hydropower plants using nanotechnology: Harnessing energy through hydro power using nanomaterials is almost similar to wind energy extraction except the method of harnessing is different.

Inference: Nanotechnology is an emerging field which is casting it's footprint in all most all facets of life. Renewable sources of energy is one of the most important fields requiring research, for the sustainable growth of mankind. As provisioning of cost effective and perennial source energy will hold the key to future progress. Nanotechnology is one of such manifest tool to provide the renewable sources of energy to all, at an appropriate cost and in reasonable quantum.

Environmental impact of wind energy

16.1   Introduction

Man has harnessed the energy in wind for thousands of years, both for sailing boats and powering wind mills at land. Of all renewable energy sources, wind power is the most mature in terms of commercial development. This energy source is interesting because of its renewability and its availability. Potential for development is huge and the world's capacity is far larger than the world's total energy consumption. The major challenges for further development are connected to economy, land usage, environment and grid capacity. The growth of renewable energy has been unprecedented over the past 25 years. Wind and solar have maintained double-digit growth rates since 2006. No other segment of the energy sector has grown this fast. Wind power is the most economic new power plant technology, due to reduced installations costs, no fuel costs and construction time of less than one year, compared to over 10 years to construct nuclear power plants. The effects of wind energy on the environment often are considered to be positive, through the production of renewable energy and the potential displacement of mining activities, air pollution and greenhouse gas emissions associated with non-renewable energy sources. As a result, a more complete understanding of the environmental and economic effects of any one energy source depends on a more complete understanding of how that energy source displaces or is displaced by other energy sources and it depends on a more complete understanding of the environmental and economic effects of all other available energy sources. This chapter provides analyses to understand those environmental effects, both positive and negative.

In recent years, the growth of capacity to generate electricity from wind energy has been extremely rapid. To the degree that wind energy reduces the need for electricity generation using other sources of energy, it can reduce the adverse environmental impacts of those sources, such as production of atmospheric and water pollution, including greenhouse gases, production of nuclear wastes, degradation of landscapes due to mining activity and damming of rivers. Generation of electricity by wind energy has the potential to reduce environmental impacts, because unlike generators that use fossil fuel, it does not result in the generation of atmospheric contaminants or thermal pollution and it has been attractive to many governments, organisations and individuals. But others have focused on adverse environmental impacts of wind energy facilities, which include visual and other impacts on humans and effects on ecosystems, including the killing of wildlife, especially birds and bats, some environmental effects of wind energy facilities, especially those concerning transportation.

Renewable energy: Presently, the known alternatives to energy production from fossil fuels are renewable and nuclear energy. There exist many social and environmental complications with nuclear energy. Renewable energy sources are desirable because of their contribution to greenhouse gas reduction and national security of energy supply without the complications of nuclear energy. Formal definitions of renewable energy sources vary from country to country, but there is broad agreement that wind energy, photovoltaic and solar thermal energy is considered to be renewables. Other sources that are often considered renewable include hydropower and biomass.

Increasing dependence on renewable energy sources is complicated by various barriers to implementation. Most renewable energy technologies face cost disadvantages in comparison with conventional energy technologies. (roads to and from the plant site) and transmission (roads and clearings for transmission lines), are common to all electricity generating facilities, others, such as their specific aesthetic impacts, are unique to wind energy facilities, at regional to global scales.

The generation of electricity from wind energy is surprisingly controversial. At first glance, obtaining electricity from a free source of energy—the wind— seems to be an optimum contribution to the nation's goal of energy independence and to solving the problem of climate warming due to greenhouse gas emissions. As with many first glances, however, a deeper inspection results in a more complicated story. How wind turbines are viewed depends to some degree on the environment and people's predilections, but not everyone considers them beautiful. Building wind energy installations with large numbers of turbines can disrupt landscapes and habitats and the rotating turbine blades sometimes kill birds and bats. Calculating how much wind energy currently displaces other, presumably less-desirable, energy sources is complicated and predicting future displacements is surrounded by uncertainties.

Although the use of wind energy has grown rapidly in the past 25 years, frequently subsidised by governments at various levels and in many countries eager to promote cleaner alternative energy sources, regulatory systems and planning processes for these projects are relatively immature in the United States and various other developing countries. At the national scale, regulation is minimal, unless the project receives federal funding and the regulations are generic for construction and management projects or are promulgated as guidelines.

The benefits of wind energy depend on the degree to which the adverse effects of other energy sources can be reduced by using wind energy instead of the other sources. Assessing those benefits is complicated. The generation of electricity by wind energy can itself have adverse effects and projecting the amount of wind-generated electricity available in the future is quite uncertain. In addition, the amount of potential displacement of other energy sources depends on characteristics of the energy market, operation of the transmission grid, capacity factor of the wind energy generators as well as that of other types of electricity generators and regulatory policies and practices affecting the production of greenhouse gases. Even if the amount of energy that wind energy displaces is small, it is clear that the nation will depend on multiple energy sources for the foreseeable future and reduction of environmental impacts will thereby require multiple approaches.

The complexity of assessing the environmental impacts of wind energy development can be organised in a three dimensional action space. These dimensional axes include spatial jurisdictions (local, state/regional and federal), timing of project stages (pre-project, construction, operational and post operational) and environmental and human impacts, each of which include their own time and space considerations.

Generation of electricity by wind energy has the potential to reduce environmental impacts caused by use of fossil fuels to generate electricity because, unlike fossil fuels, wind energy does not generate atmospheric contaminants or thermal pollution, thus being attractive to many governments, organisations and individuals. Others have focused on adverse environmental impacts of wind energy facilities, which include aesthetic and other impacts on humans and effects on ecosystems, including the killing of wildlife, especially birds and bats. Some environmental effects of wind energy facilities, especially those from transportation (roads to and from the plant site) and transmission (roads or clearings for transmission lines), are common to all electricity-generating plants, other effects, such as their aesthetic impacts, are specific to wind energy facilities.

16.2   Wind energy

Wind power is a relatively mature technology. It competes with other energy sources in terms of price, environmental effects and usability. With the exception of hydro power, wind power is closer to commercial profitability than any of the other renewable sources, though improved project economy is a vital challenge for wind power. Wind energy is widely applicable because wind resources are available in most countries. Among the renewable energy technologies, wind energy is relatively mature and many countries have resolved cost and technology challenges.

Wind is clean, free, indigenous and inexhaustible. Wind turbines do not need any type of fuel, so there are no environmental risks or degradation from the exploration, extraction, transport, shipment, processing or disposal of fuel. Not only is generation produced with zero emissions of carbon dioxide (during the operational phase) but it also does not release toxic pollutants (for example mercury) or conventional air pollutants (for example smog-forming nitrogen dioxide and acid rain-forming sulphur dioxide). Wind energy projects must be located in accordance with the *Planning Act*, the regional plans and the municipal plans. The regional plans provide general locations for wind projects and guidelines for integrating wind turbines with other land-uses. The municipal plans can include targets and desires related to wind energy and provide a more detailed basis for turbine location, number, height and appearance. Wind energy is a clean and environmentally friendly technology that produces electricity. Its renewable character and the fact it does not pollute during the operational phase makes it one of the most promising energy systems for reducing environmental problems at both global and local levels.

16.3 Environmental impacts of wind energy

Operation of wind power has zero emissions of harmful substances. It does not add to global warming, the 'fuel' is free and is quite evenly distributed around the world. The energy needed to produce and install the turbine amounts to three months of turbine production. But, as with other sources of energy, wind power does have an environmental impact. The impact on wildlife is likely low compared to other forms of human and industrial activity. However, negative impacts on certain populations of sensitive species are possible and efforts to mitigate these effects should be considered in the planning phase. Wind energy, like any other industrial activity, may cause impacts on the environment which should be analysed and mitigated.

16.3.1 Environmental benefits of wind energy

The environmental benefits of wind energy accrue through its displacement of electricity generation that uses other energy sources, thereby displacing the adverse environmental effects of those generators. Because the use of wind energy has some adverse impacts, the conclusion that a wind energy installation has net environmental benefits requires the conclusion that all of its adverse effects are less than the adverse effects of the generation that it displaces. However, the focus on the use of wind energy, it was not able to evaluate fully the effects of other energy sources. However, the facts are not fully evaluated, so-called life-cycle effects, those effects caused by the development, manufacture, resource extraction and other activities affiliated with all energy sources. Thus, in assessing environmental benefits of wind energy generation of electricity, the researchers focused on the degree to which it displaces or renders unnecessary the electricity generated by other sources and hence on the degree to which it displaces or reduces atmospheric emissions, which include greenhouse gases, mainly carbon dioxide (CO_2), oxides of nitrogen (NO_x), sulphur dioxide (SO_2) and particulate matter. This focus on benefits accruing through reduction of atmospheric emissions, especially of greenhouse-gas emissions, was adopted because those emissions are well characterised and the information is readily available. It also was adopted because much of the public discourse about the environmental benefits of wind energy focuses on its reduction of atmospheric emissions, especially greenhouse-gas emissions. The restricted focus on benefits accruing through reduction of atmospheric emissions also was adopted because the relationships between air emissions and the amount of electricity generated by specified types of electricity generating sources are well known.

However, relationships between incremental changes in electricity generation and other environmental impacts, such as those on wildlife, view sheds, or landscapes, generally are not known and are unlikely to be proportional. In addition, wind-powered generators of electricity share some kinds of adverse environmental impacts with other types of electricity generators (for example, some clearing of vegetation is required to construct either a wind energy or a coal-fired power plant and its access roads and transmission lines). Therefore, calculating the extent to which wind energy displaces other sources of electricity generation does not provide clear information on how much, or even whether, those other environmental impacts will be reduced.

Primarily, wind energy do not cause water or air emissions and do not produce any kind of hazardous waste as well. Moreover, wind power does not make use of natural resources like oil, gas or cause and therefore will not cause damage to the environment through resource transportation and extraction and also do not need consequent amounts of water during operation.

Wind energy is not only a favourable electricity generation technology that reduces emissions (of other pollutants as well as CO_2, SO_2 and NO_x), it also avoids significant amounts of external costs of conventional fossil fuel-based electricity generation. More and more use of wind energy should be made in order to prevent the problem of global warming. Wind energy plants are considered a green power technology because it has only minor impacts on the environment. Wind energy plants produce no air pollutants or greenhouse gases.

Wind energy is an ideal renewable energy because:

1. It is a pollution-free, infinitely sustainable form of energy.
2. It doesn't require fuel.

3. It doesn't create greenhouse gases.

4. It doesn't produce toxic or radioactive waste.

16.3.2 Disadvantages of environment on wind energy

Any means of energy production impacts the environment in some way and wind energy is no different. Like every other energy technology, wind power plants do have some effects on the environment. Wind turbines cause virtually no emissions during their operation and very little during their manufacture, installation, maintenance and removal. Compared to the environmental impact of traditional energy sources, the environmental impact of wind power is relatively minor.

Wind farms are often built on land that has already been impacted by land clearing. The vegetation clearing and ground disturbance required for wind farms is minimal compared with coal mines and coal-fired power stations. If wind farms are decommissioned, the landscape can be returned to its previous condition.

The major challenge to using wind as a source of power is that the wind is intermittent and it does not always blow when electricity is needed. Wind energy cannot be stored (unless batteries are used) and not all winds can be harnessed to meet the timing of electricity demands.

Good wind sites are often located in remote locations, far from cities where the electricity is needed. Wind resource development may compete with other uses for the land and those alternative uses may be more highly valued than electricity generation. Although wind power plants have relatively little impact on the environment compared to other conventional power plants, there is some concern over the noise produced by the rotor blades, aesthetic (visual) impacts and sometimes birds have been killed by flying into the rotors. Most of these problems have been resolved or greatly reduced through technological development or by properly sitting wind plants.

To the extent that we understand how, when and where wind energy development most adversely affects organisms and their habitat, it will be possible to mitigate future impacts through careful sitting decisions.

16.4   Ecological impacts of wind energy

There are two major ways that wind energy development may influence ecosystem structure and functioning—through direct impacts on individual organisms and through impacts on habitat structure and functioning. Environmental influences of wind energy facilities can propagate across a wide range of spatial scales, from the location of a single turbine to landscapes, regions and the planet and a range of temporal scales from short-term noise to long-term influences on habitat structure and influences on presence of species.

The ecological influences of wind energy facilities are complex and can vary with spatial and temporal scale, location, season, weather, ecosystem type, species and other factors. Moreover, many of the influences are likely cumulative and ecological influences can interact in complex ways at wind energy facilities and at other sites associated with changed land-use practices and other anthropogenic disturbances. Wind turbines cause fatalities of birds and bats through collision, most likely with the turbine blades.

Plants and animals throughout an ecosystem respond differently to these changes. There might also be important interactions between habitat alteration and the risk of fatalities, such as bat foraging behaviour near turbines.

Ecological aspects of wind turbines are discussed in detail in chapter 18.

Standardised studies should be conducted before sitting and construction and after construction of wind energy facilities to evaluate the potential and realised ecological impacts of wind development. Pre-sitting studies should evaluate the potential for impacts to occur and the possible cumulative impacts in the context of other sites being developed or proposed. Likely impacts could be evaluated relative to other potentially developable sites or from an absolute perspective. In addition, the studies should evaluate a selected site to determine whether alternative facility designs would reduce potential environmental impacts. Post-construction studies should focus on evaluating impacts, actual versus predicted risk, causal mechanisms of impact and potential mitigation measures to reduce risk and reclamation of disturbed sites.

16.4.1   Impacts on humans

The human impacts include aesthetic impacts, impacts on cultural resources, such as historic, sacred, archeological and recreation sites, impacts on human health and wellbeing, specifically from noise and from shadow flicker, economic and fiscal impacts and the potential for electromagnetic interference with television and radio broadcasting, cellular phones and radar.

16.4.2   Shadow flicker

As the blades of a wind turbine rotate in sunny conditions, they cast moving shadows on the ground resulting in alternating changes in light intensity. This phenomenon is termed shadow flicker. Shadow flicker is different from a related strobe-like phenomenon that is caused by intermittent chopping of the sunlight behind the rotating blades. Shadow flicker intensity is defined as the difference or variation in brightness at a given location in the presence and absence of a shadow. Shadow flicker can be a nuisance to nearby humans and its effects need to be considered during the design of a wind energy project.

In the United States, shadow flicker has not been identified as causing even a mild annoyance. In Northern Europe, on the other hand, because of the higher latitude and the lower angle of the sun, especially in winter, shadow flicker can be a problem of concern.

Shadow flicker is a function of several factors, including the location of people relative to the turbine, the wind speed and direction, the diurnal variation of sunlight, the geographic latitude of the location, the local topography and the presence of any obstructions. Shadow flicker is not important at distant sites (for example, greater than 1000 ft from a turbine) except during the morning and evening when shadows are long. However, sunlight intensity is also lower during the morning and evening, this tends to reduce the effects of shadows and shadow flicker. The speed of shadow flicker increases with wind-turbine rotor speed.

Shadow flicker may be analytically modelled and several software packages are commercially available for this purpose. An online tool for simple shadow calculations for flat topography is also available. A typical result might indicate, for example, that a house 300 meters from a 600 kW wind turbine with a rotor diameter of 40 meters will be exposed to moving shadows for approximately 17–18 hr annually, out of a total of 8760 hr in a year.

16.4.3   Public revenues and costs

Like other industries, a wind energy project generates tax dollars for the local government.

In addition, as with the private economy, the wind energy project may indirectly generate taxes for the local government. However, as discussed above with regard to the private economy, an assessment of fiscal benefits in the form of tax revenue should be based on changes that would occur if the project was built versus if it was not built. The project may encourage some forms of economic development that generate taxes, but it may deter others.

A wind energy facility also may entail public costs. Some of these, such as improvements of local public roads accessing the facility, will be obvious. Others, such as improved community services that may be expected in the wake of the development, will be indirect and less obvious. Taken together, the costs to a small, rural government have the potential to be significant.

16.4.4   Electromagnetic interference

Through Electromagnetic Interference (EMI), wind energy projects conceivably can have negative impacts on various types of signals important to human activities – television, radio, microwave/radio fixed links, cellular phones and radar. Electromagnetic interference is electromagnetic (EM) disturbance that interrupts, obstructs, or otherwise degrades or limits the effective performance of electronics or electrical equipment. It can be induced intentionally, as in some forms of electronic warfare, or unintentionally, as a result of spurious emissions and responses and intermodulation products. In relation to wind turbines, two issues are relevant: (i) possible passive interference of the wind turbines with existing radio or TV stations and (ii) possible electromagnetic emissions produced by the turbines.

There are several ways in which electromagnetic waves can deviate from their intended straightline communication paths. These include:

1. Blocking the path with an obstacle, thus creating a 'shadow' or area where the intended EM wave will not occur. To a large extent, the 'blocking' depends on the size of the obstacle as a function of the wavelength of the electromagnetic wave.
2. Refraction of the EM wave. Refraction is the turning or bending of any wave, such as a light or sound wave, when it passes from one medium into another with different refractive properties.
3. Television broadcast transmissions (approximately 50 MHz-1 GHz).
4. Radio broadcast transmissions (approximately 1.5 MHz [AM] and 100 MHz [FM]).
5. Microwave/radio 'fixed links' (approximately 3 GHz-60 GHz).
6. Mobile phones (approximately 1 or 2 GHz).
7. Radar.

Television

The main form of interference to TV transmission caused by wind energy projects is the scattering and reflection of signals by the turbines, mainly the blades. In relation to the components that make up a wind turbine, the tower and nacelle have very little effect on reception (i.e., only a small amount of blocking, reflection and diffraction occurs). This is backed up by laboratory measurements that show that the tower introduces only a small, localised (up to approximately 100 m) attenuation of the signal. The British Broadcasting Corporation has issued recommendations based on a simple concept for calculating the geometry associated with reflected signals from wind turbines and how directional receiving aerials can provide rejection of the unwanted signals.

Typical mitigation requirements include:

1. Re-orientation of existing aerials to an alternative transmitter.
2. Supply of directional aerials to mildly affected properties.
3. Switch to supply of cable or satellite television (subject to parallel broadcast of terrestrial channels).
4. Installation of a new repeater station in a location where interference can be avoided (this is more complex for digital but also less likely to be required for digital television).

Radio

Available literature indicates that effects of wind projects on both Amplitude Modulated (AM) and Frequency Modulated (FM) radio transmission systems are considered to be negligible and only apply at very small distances from the wind turbine (i.e., within tens of meters). For AM transmissions, this is due to low broadcast frequencies and long (100+ meter) signal wavelengths, which makes distortion difficult even for very large wind turbines. For FM transmissions, this is due to the fact that ordinary FM receivers are susceptible to noise interference only while operating in the threshold regions relative to signal-to-noise ratios. Thus, a distorted audio signal may be superimposed on the desired sound close to a wind turbine, potentially causing interference, only if the primary FM signal is weak.

Cellular phones

Mobile-phone reception depends greatly on the position of the mobile receiver. Therefore, the movement of the receiver and the topography—including both natural and unnatural obstacles—have a major impact on the quality of the signal. The mere movement of the receiver can ensure that wind turbines will have a very minimal effect, if any, on communication quality.

Radar

The potential for interference of wind turbines with radar is only partially understood. If there is such interference, it would primarily affect military and civilian air-traffic control. In addition, NWS weather radars might be affected.

One finding concludes that 'wind farms located within radar line of sight of air defense radar have the potential to degrade the ability of the radar to perform its intended function. The magnitude of the impact will depend upon the number and locations of the turbines. Should the impact prove sufficient to degrade the ability of the radar to unambiguously detect objects of interest by primary radar alone this will negatively influence the ability of military forces to defend the nation.'

16.4.5   Impacts on humans

Wind energy facilities create both positive and negative recreational impacts. On the pospaitive side, many windenergy projects are listed as tourist sights: some offer tours or provide information areas about the facility and wind energy in general and several are considering incorporating visitor centers. There are two types of potential negative impacts on recreational opportunities–direct and indirect. Direct impacts can result when existing recreational activities are either precluded or require rerouting around a wind energy facility. Indirect impacts include aesthetic impacts (addressed above) that may affect the recreational experience. These impacts can occur when scenic or natural values are critical to the recreational experience. In analysing impacts on historic, sacred and archeological sites, the primary concern is that no permanent harm should be done that would affect the integrity of the site. Whether or not a wind energy project would damage the resource may depend on the specific nature of the historic resources involved. Unlike housing developments, wind energy projects cannot be screened from view, except behind intervening topography and vegetation. Such issues are likely to arise as wind projects are proposed in cultural landscapes and guidance as to what constitutes an undue impact to historic or sacred sites and areas will be necessary.

16.4.6 Impacts on human health

Wind energy projects can have positive as well as negative impacts on human health and wellbeing. The positive impacts accrue mainly through improvements in air quality, as discussed previously in this chapter. These positive impacts (i.e., benefits) to health and well-being are diffuse, they are experienced by people living in areas where conventional methods of electricity generation are used less because wind energy can be substituted in the regional market. In contrast, to the extent that wind energy projects create negative impacts on human health and wellbeing, the impacts are experienced mainly by people living near wind turbines that are affected by noise and shadow flicker.

16.4.7 Local economic impacts

Wind energy projects can have a range of economic and fiscal impacts, both positive and negative. Some of those impacts are experienced at the national or regional level, these involve and for example, tax credits and other monetary incentives to encourage wind energy production, as well as effects of wind energy on regional energy pricing.

16.4.8 Visual impact

Landscape perceptions and visual impacts are key environmental issues in determining wind farm applications related to wind energy development as landscape and visual impacts are by nature subjective and changing over time and location.

The characteristics of wind developments may cause landscape and visual effects. These characteristics include the turbines (size, height, number, material and colour), access and site tracks, substation buildings, compounds, grid connection, anemometer masts and transmission lines. Another characteristic of wind farms is that they are not permanent, so the area where the wind farm has been located can return to its original condition after the decommissioning phase. While visual impact is very specific to the site at a particular wind farm, several characteristics in the design and sitting of wind farms have been identified to minimise their potential visual impact.

16.4.9 Noise impact

Noise from wind developments has been one of the most studied environmental impacts of this technology. Noise, compared to landscape and visual impacts, can be measured and predicted fairly easily. Noise from wind energy and its control are discussed in detail in chapter 17.

16.4.10 Impact on land use

National authorities consider the development of wind farms in their planning policies for wind energy projects. Decisions on sitting should be made with consideration to other land users. Regional and local land-use planners must decide whether a project is compatible with existing and planned adjacent uses, whether it will modify negatively the overall character of the surrounding area, whether it will disrupt established communities and whether it will be integrated into the existing landscape. Land use planning rules in some countries recommend avoiding areas with protected designations, in others, specific areas have been earmarked for potential wind farm development.

16.5   Reducing the negative environmental impacts of wind energy

The negative environmental impacts from wind energy installations are much lower in intensity than those produced by conventional energies, but they still have to assessed and mitigated when necessary. There are specific conditions that must be in place before an area can be considered suitable for a wind farm development. These conditions include factors such as wind climate, topographical, logistical and ecological constraints. From a technical and economic standpoint, the most mature form of renewable and 'clean' energy is wind energy. It can effectively contribute to combating climate change while at the same time providing various environmental, social and economic benefits. On the other hand, it is necessary to minimise the impact of the wind energy, particularly in terms of environment (preservation of protected areas) and human health (noise and visual impact).

16.6   Environmental impacts of different technologies

16.6.1 Environmental impacts of wind power

A wind farm, when installed on agricultural land, has one of the lowest environmental impacts of all energy sources. It occupies less land area per kilowatt-hour (kWh) of electricity generated than any other energy conversion system, apart from rooftop solar energy and is compatible with grazing and crops, it generates the energy used in its construction in just 3 months of operation, yet its operational lifetime is 20–25 years, greenhouse gas emissions and air pollution produced by its construction are very tiny and declining.

There are no emissions or pollution produced by its operation, in substituting for base-load (mostly coal power) wind power produces a net decrease in greenhouse gas emissions and air pollution and a net increase in biodiversity, modern wind turbines are almost silent and rotate so slowly (in terms of revolutions per minute) that they are rarely a hazard to birds.

Modern wind turbine designs have significantly reduced the noise from turbines. Turbine designers are working to minimise noise, as noise reflects lost energy and output. Noise levels at nearby residences are managed through the siting of turbines, the approvals process for wind farms and operational management of the wind farm. The noise limit for wind farms is 35 A-weighted decibels, which is usually around 5 Aweighted decibels above a quiet countryside. Alternatively, the limit is 5 A-weighted decibels above the level of background noise (i.e., without wind farm noise), if that is greater than 35 A-weighted decibels. Low frequency sound and infrasound (i.e., usually beneath the threshold of human hearing) are everywhere in the environment. They are emitted from natural sources such as wind and rivers and artificial sources such as traffic and air conditioning. Modern turbine designs which locate the blades upwind instead of downwind have significantly reduced the level of infrasound. Scientific and health authorities have found the low level of infrasound emitted by wind turbines pose no health risks.

Wind turbines may create shadow flicker on nearby residences when the sun passes behind the turbine. However, this can easily be avoided by locating the wind farm to avoid unacceptable shadow flicker, or turning the turbine off for the few minutes of the day when the sun is at the angle that causes flicker. Shadow flicker is considered in the NSW development assessment process to ensure potential impacts are addressed. Many energy policy studies have noted how wind turbines present direct and indirect hazards to birds, other avian species and bats. Birds can directly smash into moving or even stationary turbine blades, crash into towers and nacelles and collide with local distribution lines. These risks are exacerbated when turbines are placed on ridges and upwind slopes or built close to migration routes. Some species, such as bats, face additional risks from the rapid reduction in air pressure near turbine blades, which can cause internal hemorrhaging.

Comparative assessment of avian mortality for fossil fuel, nuclear and wind power plants in the United States is presented in. It was stated that for wind turbines, the risk appears to be greatest to birds striking towers or turbine blades. For fossil-fuelled power stations, the most significant fatalities come from climate change, which is altering weather patterns and destroying habitats that birds depend on. For nuclear power plants, the risk is almost equally spread across hazardous pollution at uranium mine sites and collisions with draft cooling structures. Yet, taken together, fossil-fuelled facilities are about 17 times more dangerous to birds on a per GWh basis than wind and nuclear power stations. Timely decommissioning of turbines that are no longer in use is a standard condition of consent for wind farms in NSW. Decommissioning wind farms is a straightforward task. In Europe many older wind farms are being re-powered with new turbines, this could also be expected to occur in New South Wales. Existing wind sites have considerable value – the wind resource is well understood and structural foundations, electricity transmission and local community acceptance are already in place.

Noise from wind energy and its control

17.1 Introduction

Noise is defined as unwanted sound and can range from high levels (causing physical damage) to levels only just audible. Noise is measured in decibels (dB) and human hearing ranges from approximately 0 dB (the threshold of hearing) to 120 dB (the threshold of pain). A change of 3 dB in noise level is just perceptible under normal circumstances and results from doubling or halving the number of noise sources. A change of 10 dB corresponds to an approximate doubling of perceived loudness. The way a noise sounds is partially determined by its pitch or frequency, which is measured in Hertz (Hz). Human hearing covers frequencies from 20 Hz to 20000 Hz and is less sensitive at low and high frequencies. Because the ear is less sensitive at low and high frequencies, this has to be allowed for in noise measurements by reducing the response of measurement equipment by a similar amount. Various 'weightings' to measurements of sound pressure level are available for this purpose and the 'A' weighting is the one which is normally used for environmental noise measurement, as it appears to correlate best with human subjective response, with the resulting weighted noise levels referred to as being in dB(A). Sounds which cover a range of only a few Hz are referred to as being tonal. Sounds which cover a wide range of frequencies do not have any tonal character and are referred to as being broadband, i.e., not dominated by any particular frequency.

Most noise is not steady but varies over time. As a result, statistical averaging is used to quantify noise levels over a given time period. For instance, the $L_{Aeq,T}$ is a measure of the average noise over a 'T' measurement period such as $L_{Aeq,10min}$ represents the ('A' weighted) average noise level over a 10 minute period. Percentiles are typically used to describe noise level over a percentage of the measurement period such as the $L_{A90,T}$ which is the noise level exceeded for 90% of a 'T' measurement period and $L_{A10,T}$ which is the noise level exceeded for 10% of a 'T' measurement period. These can be used to express a range of descriptive percentiles from L_{A1} to L_{A99}, $L_{A90,T}$ is normally used to describe 'background noise'. It is a requirement of the environmental protection agency to ensure that licensed sites (IPPCi licence and waste licence) do not cause unacceptable impact on the environment which includes noise impact on the human environment. IPPC and waste licences contain several conditions in relation to noise including separate daytime and night-time noise limits, monitoring and reporting protocols and site specific noise monitoring locations. It is the responsibility of each licensee to ensure that the site is compliant with these conditions and does not have a detrimental effect on the local noise climate. IPPC and waste licences may also contain conditions to control site noise and the resultant noise impact. The overriding objective is to minimise or negate noise impact at Noise Sensitive Locations (NSLs).

17.2 Noise emissions from wind turbines

17.2.1 Mechanical noise

Like any piece of equipment containing moving parts, a wind turbine emits a certain amount of mechanical noise. This is generally dominated by noise from the gearbox (although not all turbines have gearing systems) and to a lesser extent, the generator. There may also be noise from cooling fans, oil pumps and other auxiliary equipment. In addition, the yaw motors make occasional noise as they turn the turbine to face into the wind. As for all rotating machinery, associated mechanical noise may have tonal components which generate noise with a specific pitch dependant on the speed of rotation. Where it occurs it can have a significant effect on the way the resultant noise is perceived. A noise with a significant tonal component is likely to give the impression of a noise which is 5 dB or louder than a noise of the same level without any tonal component.

Modern wind turbine design incorporates insulation in the nacelle to prevent airborne transmission of mechanical noise. Nacelle isolation is also incorporated to prevent vibration from moving parts being transmitted into the tower or blade assemblies and subsequently re-radiated as structure-borne noise.

This has become the norm over 20 years of commercial wind turbine operation in Ireland such that it is now unusual for *tonal components* to be a significant factor in wind turbine noise. However, tonal components continue to be regulated by planning conditions which include a penalty for such effects. Standard ISO 1996 'Acoustics' – Description, measurement and assessment of environmental noise Part 2:2007 Determination of environmental noise levels provides for tonal assessment and penalties. Tonal components should likewise be covered within manufacturers test data and warranty documents. It should be noted that typical Agency licences prohibit tonal noise at night and apply a penalty during the day.

17.3 Noise reduction

17.3.1 Mechanical noise reduction

One source of mechanical noise is vibrations induced by the rotating components. Vibration control is used to suppress or eliminate unwanted vibrations. Depending on the case, different control laws can be chosen in order to minimise unwanted vibration. Additionally, damping or increasing the effective mass can be realised by the controller. Inferentially, the absorber works as an active system. This includes the use of sound isolating materials, insulation and closing the holes in the nacelles which would decrease the sound transmitted to the air. Aside from loss of power and increased maintenance costs, faulty gearboxes also increase noise levels in wind turbines. As a result, researchers are developing fault diagnostic systems for gearboxes with applications to wind turbines. A system has been developed to integrate singular value decomposition noise reduction, time-frequency analysis and order analysis methods in order to identify weak faults objectively and effectively. More recently, efforts have been taken to develop intelligent techniques for online condition monitoring in machinery systems such as wind turbines. Several neural fuzzy classification techniques have been proposed for fault detection.

17.3.2   Aerodynamic noise

Although rotational speed of modern wind turbines is relatively slow at around 20 revolutions per minute, the speed at which the tips of the blade travel can be fairly high due to the length of the blades. The equivalent tip speed for a 90 meter blade diameter turbine would be 340 km/hr; approaching the speed of sound. As a result, aerodynamic noise from large wind turbines can be fairly significant and, due to improvements in the control of mechanical noise, is now the more dominant noise source. The sound is broadband and may be likened to the noise of rushing water. It can be controlled to a certain extent by careful blade design to minimise vortex shedding from the blade tips and trailing edge noise where the flow over the two sides of the blades interacts. It is inevitable that there will be a certain amount of residual noise from where the boundary layer separates from the blade surface and from where the turbulent incoming airflow meets the leading edge of the blade.

17.3.3   Aerodynamic noise reduction

Adaptive approaches

There are a number of adaptive noise reduction approaches for aerodynamic noise, including varying the speed of rotation of the blades. Since an increase in rotational speed will also lead to increased noise production, lowering the rotational speed will lead to decreased sound. However the rotational speed decrease reduces power output and therefore should only be implemented within a certain range of wind velocities, since high winds also have the added benefit of masking the sound of the wind turbine with the sound of the wind itself. The pitch angle of the wind turbine blades also has an important role in noise production. An increase in pitch angle will lead to a reduction in the angle of attack. As the angle of attack increases, the size of the turbulent boundary layer on the suction side of the airfoil grows, thereby increasing noise production in the wind turbine. Therefore, if the pitch angle is increased, a thinner boundary layer results on the suction side, which is considered the strongest source of noise production. This also implies that, on the pressure side, the effect is the opposite; therefore when using this method for noise control, it is important to find the appropriate pitch angle range for optimal noise control. As with the previous method, the major drawback to this adaptive noise control method is the corresponding reduction of power since the angle of attack is decreased. Despite a potential loss in power due to pitch angle changes that reduce noise, more wind turbines can potentially be built within a specified area.

Wind turbine blade modification methods

The main drawback to adaptive methods (an overall reduction of power) is a hindrance to that method of noise control. By breaking down the noise sources it can be seen that the maximum noise contribution occurs within the trailing edge. Different aeroacoustic prediction models such as semi-empirical airfoil self-noise models and simplified theoretical models are available in the literature. Kamruzzaman and others presented results using a well-known TNO-Blake model with Computational Aeroacoustics (CAA) and validated the predictions against wind tunnel measurements. The predictions are based on Reynolds Averaged Navier-Stokes simulation results. Such predictions are beneficial in determining the region with the highest aeroacoustic levels. Researchers established that the region between about 75–95% span is exposed to the maximum flow velocities and it has the highest aero-acoustic noise levels. Furthermore, experimental tests showed that most of the noise is generated when the blade is moving downwards in a clockwise rotation. Therefore, the majority of modification procedures are aimed at reducing the noise in this area. Research had been conducted into the use of aero-acoustically optimised airfoils and trailing edge serration, while trailing edge brushes have also been considered to control the noise generated from this portion of wind turbine blades.

17.3.4   Noise characteristics

General characteristics

In general, noise from wind turbines increases with wind speed and rotational speed. Most modern wind turbines are pitch regulated variable speed turbines which have a characteristic noise profile of steeply increasing noise with wind speed up to the point at which the turbine is generating its 'rated power' or maximum power above which there is usually no increase in noise. For 'stall regulated' machines, noise can increase considerably above rated power of the turbine and such designs are becoming less popular as a result.

Spectral characteristics

The aerodynamic noise is broadband and spread across the audible frequency range. There is a common misconception that there is a significant component of low frequency noise but this is not the case. As distances increases from a noise source, the noise spectrum becomes more biased towards the low frequencies. This is caused by greater attenuation of middle to high frequencies by atmospheric effects, with reduced attenuation of low frequencies.

This is true of any broadband noise such as road traffic or the sound of the sea. This may, therefore, be a significant characteristic for a large wind farm site when heard from a distance although close to the turbines, it would not be significant.

There is similarly no significant infrasound from wind turbines. Infrasound is high level sound at frequencies below 20 Hz. This was a prominent feature of passive yaw 'downwind turbines' where the blades were positioned downwind of the tower which resulted in a characteristic 'thump' as each blade passed through the wake caused by the turbine tower. With modern active yaw turbines (i.e., the blades are upwind of the tower and the turbine is turned to face into the wind by a wind direction sensor on the nacelle activating a yaw motor) this is no longer a significant feature.

Tonal content in the noise, where it occurs, can be identified by peaks in the frequency spectrum and quantified according to recognised procedures dealing with the audibility of tonal components in broadband noise such as ISO 1996 Part 2:2007. As before, it is uncommon for wind turbine noise to contain tonal components but these can be minimised through insulation and isolation as discussed above.

Temporal characteristics

Wind turbine noise fluctuates at a rate depending on the speed of rotation. Technically this is known as amplitude modulation, which means variation in noise level, although it is sometimes referred to as aerodynamic modulation or blade swish. This feature of amplitude modulation can readily be observed close to a single wind turbine such that on the downward stroke a different level and timbre of noise may be heard to that as it passes the tower, or as occurs on its upward stroke. As distance from the turbine increases this effect reduces although for some configurations of turbine sites the effect has been noted to be audible at residential locations.

The mechanism behind this is not clearly understood. However, features which are thought to enhance this effect are:

1. Close spacing of turbines in linear rows.
2. Tower height to rotor diameter ratio less than approximately 0.75.
3. Stable atmospheric conditions.
4. Topography leading to different wind directions being seen by the blades at different points in their rotation.

It is difficult to quantify the degree of amplitude modulation even though it is possible to identify the effect from examination of a graphical display of variation in noise level. Evaluation of the significance of such effects is not covered by any recognised process. Because such effects may, like tonal noise, give the impression of a noise which is 5 dB or more louder than a noise of the same level without any such components, methods are being developed using complex signal processing to allow such evaluation to be repeatedly and consistently carried out such that appropriate corrections can be applied where necessary for regulatory purposes.

17.4   Subjective response to wind turbine noise

Human subjective response to noise depends on a number of factors as well as overall noise level, including the characteristics of the noise, the duration and time of exposure, the activity being carried out during exposure to the noise, the expectations which the person has of their acoustic environment, the level of noise from other sources, hearing sensitivity and non-acoustic influences, such as attitude towards the noise and general health. It is well established that people respond differently to different types of noise. Different individuals will also respond differently to the same type of noise. In general people are prepared to tolerate higher levels of road traffic noise than noise from commercial installations or premises.

Traditionally, wind turbines tend to be installed in areas of low background noise and, particularly at night, it is often the case that there may be no other significant sources of noise other than the noise of the wind as it blows through trees and other foliage. Additionally, people often chose to live in rural areas specifically because they value the lack of noise from man made sources, so any such noise which is audible may be a source of disturbance/ annoyance and possibly complaint. Although tonal noise is no longer a significant issue from turbines, excessive amplitude modulation can attract additional attention, particularly if this can be heard whilst trying to get to sleep either at the start of the night or when a person has been woken up by other causes. Although the level of noise generated internally, even with windows open, is usually insufficient to cause sleep disturbance, the stress it may generate, even if only just audible, may be sufficient to extend the time required to fall asleep. This is likely to be exacerbated by excessive audible amplitude modulation, the level of which itself may vary with time, where it draws additional attention to the noise. Wind turbines installed on EPA licensed sites, many of which are located in mixed residential and industrial areas rather than quiet rural locations, are likely to be perceived as an addition to the plant and equipment which is already in use at such sites rather than a completely new noise source. As a result, it is possible that such installations may be accepted more readily as they will not be perceived to change the character and acoustic environment of the area as a completely new development might.

17.5   Potential noise impacts of wind turbine operations

17.5.1   Operational noise impacts

Noise is generated by wind turbines as they rotate to generate power. This only occurs above the 'cut-in' wind speed and below the 'cut-out' wind speed. Below the cut-in wind speed there is insufficient strength in the wind to generate efficiently and above the cut-out wind speed the turbine is automatically shut down to prevent any malfunctions from occurring. The cut-in speed at turbine hub height is normally between 3 and 5 meters per second (m/s) and the cut out wind speed is normally around 25 m/s.

Noise levels are greater when the wind is blowing from the turbines towards the receiver location. With cross-winds (where the wind blows across a path between the turbine and the receiver), noise levels can be expected to be around 2 dB lower than downwind noise levels. For upwind propagation, (i.e., where the wind blows from the receiver towards the turbine), the noise level can be expected to be at least 10 dB lower. Exception to this 10 dB reduction occur within a distance of up to five times the hub height or where the ground falls away rapidly between the turbine and the receiver.

Although wind turbines are often situated in rural locations where there is little other man-made noise, turbine noise can be masked by wind as it blows through trees and other foliage. This can be very helpful in reducing the impact of turbine noise, particularly for variable speed turbines at low wind speeds. The lower noise output at low wind speeds is masked by the low wind generated background noise. Caution should, however, be applied in any assumptions made about turbine noise and masking from wind effects. Due to the heights of modern commercial wind turbines and 'stable atmospheric' conditions during the night-time hours, the wind speed at ground level may be low but wind speed at hub height may be sufficient to generate significant noise levels. This is sometimes referred to as the 'wind shear' effect. As a result of this effect, it is important to quantify background noise in terms of measured or derived hub height wind speed. Site specific wind shear can be calculated by analysis of wind speed data collected at different meteorological wind vane heights and applied to generate hub height derived wind speed data. It can be acceptable to convert the hub height wind speeds to standardised 10 meter height wind speeds for comparison with manufacturers noise data without introducing any additional wind shear effects.

Modelling of turbine noise impacts

At any given time turbine noise impact depends on wind speed, wind direction and background noise. The only recognised methodology for aggregating the impact as it varies with wind conditions would be to calculate a yearly L_{Aeq}, or yearly L_{den} and L_{night} values in line with the European Environmental Noise Directive 2002/49/EC13. There is no conclusive evidence that this correlates better with perceived impact than worst case conditions although there will inevitably be a mitigating factor of how often such worst case conditions occur. Noise impact is therefore usually quantified as a worst case compared to either a maximum noise limit or noise limits derived for different wind speeds.

17.5.2 Noise sensitive locations

Noise Sensitive Locations (NSLs) are deemed to be any location in which the inhabitants can be disturbed by noise from the site (including turbine noise). This incorporates the previous definition for a NSL provided in the previous agency guidance notes which also covers the definition for a NSL provided in the Department of the Environment, Heritage and Local Governments (DoEHLGs) 2006 guidance document 'Wind Farm Planning Guidelines'.

17.6 Noise limits

Noise limits are normally set such that the majority of the population are not disturbed by the noise levels specified. It is therefore inevitable, that some people may be annoyed by the permitted noise levels. As noise level decreases, the likelihood of disturbance decreases. It is well established that noise which is attributable to an individual, company or other organisation causes more annoyance, in general, than noise from a source which is not attributable to any specific person or group such as road traffic. Noise from individuals, such as neighbours or other community sources, tends to cause the most annoyance.

17.6.1 Review of noise limits in existing EPA noise guidance

The second EPA Noise Guidance Note (NG2) 'Guidance Note for Noise in Relation to Scheduled Activities' contains generally applicable noise limits of 55 dB LAr,T for daytime (08:00–22:00) and 45 dB LAeq for night-time

(22:00–08:00). Tonal and impulsive noise should be avoided during daytime hours but the daytime LAeq,T can be corrected arithmetically for the presence of tonal and impulsive noise as LAr,T. At night there should be no tonal or impulsive components. It is noted that in particularly quiet areas where the background noise levels are below approximately 35 dB L90 (likely to be LA90), lower noise limits may be more appropriate.

17.6.2   Different approaches in relation to wind turbine noise limits

Noise assessment is usually based on comparison of predicted external noise levels with established criteria. Such criteria may be relative, absolute, or a combination of the two.

Relative criteria

Relative criteria are based on levels of existing noise and may be set either:

1. According to the change in a given noise index (for instance, increase in road traffic noise is quantified in terms of a change in L_{Aeq} or L_{A10}).
2. According to a new level of sound in one index relative to the existing level of sound using a different index (for instance where the L_{Aeq} of new noise is compared to the L_{A90} of existing background noise).

Absolute criteria

Absolute criteria are fixed and independent of background noise but may vary according to type of area (urban, sub-urban, rural residential) and time of day (day, evening and night).

Hybrid criteria

The typical limits used to assess wind turbine noise are a combination of relative and absolute criteria.

17.7   Noise mitigation of wind turbines

It may be necessary to use 'noise constrained' versions of wind turbines at some sites. This noise constraining means that the turbines maximum rotational speed and hence its power output, is reduced to a value below that which would apply in the non-constrained version. The fact that the turbine cannot operate up to its rated power should not be seen as a negative point necessarily but the potential power output of the turbine, under such conditions, would need to be taken into account in the planning balance. It may only be necessary to constrain the wind turbine for specific wind speeds in specific directions. This would be suitable where receivers are predominantly in one or two directions from the site and the turbine has been predicted to generate noise levels above the applicable limits at these locations. This would allow the turbine to operate optimally in all other directions where impact had not been determined. This approach is normally implemented within the control software which governs the turbine operation and can be pre-programmed and requires no additional input from any operator.

17.8   Noise impact assessment methodology

The following noise impact assessment methodology is described in order to assess predicted cumulative site noise and compliance with either fixed or background noise derived limits:

1. Predicting noise levels of the proposed turbine(s).
2. Assessing whether the cumulative predicted turbine and site noise is compliance with the fixed limits as per this guidance note (NG3).
3. If the predicted noise levels are not compliant with the fixed limits, background noise limits can be derived and the cumulative predicted noise assessed against these.

17.8.1   Noise modelling of proposed turbines

In carrying out an assessment prior to development taking place, certain assumptions are necessary to allow predicted noise levels to represent typical conditions likely to occur in practice. These may include the source sound power levels of equipment to be installed, the ground and meteorological conditions and the way the equipment is likely to operate in practice. Allowances are usually made to ensure that noise limits are not exceeded by likely worst case scenarios. The following procedures should be followed:

1. Turbine noise predictions should be carried out for wind speeds from cut-in (3–5 m/s) up to a standardised 10 meter height wind speed of 12 m/s. Sound power data is rarely supplied above this wind speed.
2. The predictions should be carried out using the methodology defined in ISO 9613-2, Acoustics – Attenuation of Sound during Propagation Outdoors – Part 2: General method of calculation applying the following modelling parameters:
   (a) The ground coefficient should be modelled as $G = 0.5$, as recommended in prediction and measurement of wind turbine noise, except where transmission will be over completely hard ground or water when a value of $G = 0$ should be used.
   (b) The assumed sound power levels should be those guaranteed by the manufacturer of the proposed turbine and should be backed up by a warranty statement.

(c) The assumed receiver height should be 4 meters even if the receiver is a one storey property.

3. Predictions should be carried out using an assumed octave band noise spectrum taken from a turbine test report for each wind speed where available, or for the reference wind speed of 8 m/s, if not.

4. Attenuation for topographical or other shielding should only be applied where the whole of the turbine is out of line of site, when an attenuation of 2 dB may be assumed.

17.8.2 Assessment of predicted cumulative site noise against fixed limits

Following prediction of the turbine noise, the cumulative rated site noise level should be predicted by adding together the existing rated site noise and the predicted turbine noise levels. Existing site noise should be provided by a verified model. Existing noise levels can also be provided by long-term measurements (at least two weeks) subject to prior agreement with the Agency on this approach.

17.9 Measurement equipment

Long-term noise surveys should be completed using a Type 1 Sound Level Meter (SLM) with an outdoor microphone kit. The outdoor microphone kit should allow for the meter to be stored in a lockable weatherproof box to prevent intentional tampering or weather affecting the meter. The microphone should be protected with a specific long-term microphone wind shield, following the guidance contained in 'Noise Measurements in Windy Conditions' (ETSU W/13/00386/REP). The SLM should have a recent traceable calibration either annual or biennial. The instrument should be field-calibrated prior to commencing the survey using the recommended calibration procedure and a calibrator of a known level. The meter should be checked on completion of the survey to record drift during the course of the monitoring period. The calibrator should have a recent traceable, annual calibration.

The sound level meter should be set to a frequency weighting of 'A' in accordance with international standard IEC 61672:2003 and various national standards relating to the measurement of sound pressure level representative of human hearing. The meter should be capable of recording L_{Aeq}, L_{Amax}, L_{Amin} and L_{An} statistical parameters. The meter should be capable of logging the necessary parameters on a ten minute interval and have sufficient memory capacity to store the data. Another option is to connect the meter, via., a sim card modem which allows downloading of the data remotely. This also allows for checking on the correct operation of the meter remotely.

Sufficient power should be available to the meter, either through heavy duty batteries or a solar panel. If lead acid batteries are used, health and safety risk assessment should be undertaken where these are handled or transported.

Where there is a possibility that the combined noise contains tonal or impulsive components, audio recording should be included to enable subsequent spectral and temporal analysis to be carried out. This is required to derive the associated corrections to the measured noise levels for comparison against the day-time limit or to ensure that the night-time requirements are being met. Any analysis of tonal components should follow the provisions of ISO 1996 Part 2:2007.

17.10   Meteorological measurements

For background noise measurements, it is necessary to record wind speed on site, in 10-minute periods concurrent with the noise measurements. Measurements should either be at hub height or at two lower heights (on a meteorological mast) to allow estimation of hub height wind speed. If a dedicated meteorological mast is not available, wind sensors should be mounted in accordance with the National Oceanic and Atmospheric Administrations guidance 'Guidelines for Meteorological Station Reconnaissance and Meteorological Sensor Height Measurements'. For measurements of turbine noise, meteorological data should be taken from the wind turbine nacelle anemometer(s), suitably corrected for the presence of the rotating blades and yaw data.

Rainfall should, however, be measured at a location representative of the noise measurement location(s).

17.10.1 Noise from other sources

As the long-term measurements are unattended, the meter should be set up in a position that accurately reflects the noise levels from the site (and the turbine if operating and being monitored). A suitable location should be set up away from non-site, extraneous noise sources that could influence the overall noise levels. One of the biggest problems in measuring noise from wind turbines is that of noise from other wind induced sources such as trees and foliage. It may be necessary to arrange for the turbine(s) to be shut down for certain periods to allow the contribution from other sources to be quantified and subtracted from the total measured noise to allow the turbine noise to be evaluated.

17.10.2 Appropriate curtailing of wind turbine operations when designated noise limits may be exceeded

Predicted turbine noise levels may show potential for noise limit exceedances at certain speeds or wind directions at modelled NSLs. If this occurs, the agency can set specific conditions that the turbine operations are limited for specific wind speeds or directions. If such mitigation is required, the control software should provide for a logging system to record and archive when the turbine(s) were constrained. Any proposed noise mitigation will normally be imposed through licence conditions.

17.11   Wind turbine noise control concepts

This section will outline methods of controlling both broadband swish and BTI noise.

17.11.1 Passive control methods

The most efficient means of controlling trailing edge noise is to reduce the strength of its source. One of the most direct methods for doing this is to alter the blade shape in order to influence the nature of the turbulent boundary layer at the trailing edge. Methods of doing this vary between adhoc design changes to computationally demanding aeroacoustic shape optimisation. Recently, Jones and others developed an optimisation procedure using a semi-empirical model of trailing edge noise to develop new, low noise airfoil designs. One design achieved a 2.9 dB OASPL noise reduction (over the original NACA 0012 shape used to start the optimisation process) whilst also reducing drag. It can be expected that much quieter airfoil designs will be developed as noise prediction methods become more accurate and efficient. Another important passive noise control technique for trailing edge noise is the use of trailing edge serrations.

These are saw-tooth extensions placed on the trailing edge. As originally pointed out by Howe, the serrations present a trailing edge at an angle to the stream wise flow direction thus reducing the efficiency of the edge sound source. Theoretically, serrations are able to reduce noise by a large amount. However, in practice, serrations do not reduce noise as much as theory suggests and this may be due to the production or re-orientation of turbulence by the serrations themselves. Porous trailing edge inserts are also promising noise reducing devices, but may have limited applicability due to dirt accumulation in the pores, requiring regular costly maintenance.

While shape modifications or inserts may provide an effective means of trailing edge (broadband swish) noise control, passive means of BTI noise are limited. One answer is to increase the distance between the rotor tip and tower. The current spacing between the rotor tip and tower has probably been maximised by the manufacturer. Increasing this distance will require extensive redesign of the gearbox and nacelle and could introduce more problems such as shortened mechanical life, vibration and noise.

17.11.2 Active control concepts

Swish and BTI reinforcement occurs due to in-phase noise production on multiple wind turbines. As each turbine rotates in the same direction and experiences close to the same wind speed and direction they will turn at very nearly the same angular velocity. If the azimuthal phase of a group of wind turbines is nearly the same, then we would expect that their sound would be produced at nearly the same time and propagate in a similar manner. Given that broadband swish has a forward propagating directivity, then zones of high amplitude modulation of trailing edge noise are expected. BTI noise has the directivity of a dipole, hence an array of in-phase BTI sources will create alternate zones of reinforcement and cancellation.

Active phase desynchronisation is a concept that can potentially alleviate this situation. By monitoring the phase of each blade in a wind farm, small adjustments to the rotor blade pitch or brake can be made to alter the blade's phase and ensure that noise reinforcement does not occur at a particular receiver location or locations, such as homes. While this seems a simple and cost effective solution to the problem, it may be difficult to implement without more knowledge of how the noise sources are produced, their strengths and how they propagate in the atmosphere.

17.12   Summary and conclusion

17.12.1 Sources of wind turbine noise

Modern large turbines

Mechanical noise from nacelle:
1. Mainly caused by gearbox and generator.
2. May generate tones and low-frequency noise.
3. Gearless turbines quieter than geared turbines.

Aerodynamic noise from blades:
1. Usually dominant noise source.
2. Inflow turbulence noise.
3. Tip noise.
4. Airfoil self noise $\rightarrow$ trailing edge noise.

17.12.2 Conclusion

Wind turbines have many potential noise sources: Mechanical noise, tip noise, inflow noise, airfoil self-noise.

Many noise sources can be suppressed by good design: For modern turbines the dominant noise source is trailing edge noise from the outer part of the blades.

Several design options for further noise reduction:

1. Blade add-ons.
2. Smart control strategies.
3. Blade shape (planform, airfoils, tip design).

Ecological aspects of wind energy

18.1 Introduction

The wind farms tend to have variable effects on bird populations, which can be species-, season and/or site-specific. The impacts include collision fatalities, habitat loss and disturbance resulting in displacement. The main factors that contribute to collision fatalities are proximity to areas of high bird density or frequency of movements (migration routes, staging areas, wintering areas), bird species (some are more prone to collision or displacement than others), landscape features that concentrate bird movement and poor weather conditions. In many instances, the numbers of carcasses reported are likely to be underestimates, as they are often based only on found carcasses, without accounting for scavenging and searcher efficiency.

Habitat loss as a result of wind farm construction seems to have a minor impact on birds, as typically only 2–5% of the total wind farm area is taken up by turbines, buildings and roads. However, the cumulative loss of sensitive or rare habitats may be significant, especially if multiple large developments are sited at locations of high bird use. Disturbance of birds as a result of wind farm development may arise from increased activity of people at the site and/or the presence, motion and noise of turbines. The level of disturbance to birds has been shown to vary, depending on the availability of alternative feeding or breeding habitat.

There is little relationship between the scale of a wind farm and the amount of bird mortality that has occurred. A large, appropriately sited wind farm may kill fewer birds than a small, poorly sited one. Considered in isolation, it is unlikely that small numbers of fatalities per year at a wind farm would be considered significant, unless some of those fatalities were of threatened species, in which case impacts might occur at the population level (although it should be noted that cumulative effects of small numbers of fatalities at two or more wind farms may be sufficient to result in population impacts). In contrast, a large facility may kill many birds in total, thus impacting at the population level, especially when threatened species are involved. Even relatively small increases in mortality rates may be significant for populations of some birds, especially long-lived species with generally low annual productivity and slow maturity and particularly when already rare, e.g., blue duck (*Hymenolaimus malacorhynchos*) and kaka (*Nestor meridionalis*). When considering potential impact, it is important to consider the average effect of each turbine, the cumulative effect of the total number of turbines and associated structures on a farm and even the cumulative impact of other wind farms in the range of a bird population, particularly where rare or threatened species are concerned.

As the area of the farm increases (density of turbines remaining constant), the potential for adverse effects, other than fatalities, also increases. Large facilities may cause more bird habitat to be lost or compromised, so that for ageing and breeding birds may be more inclined to avoid the area. Even in New Zealand, a large wind farm can occupy many square kilometres in area, e.g., Hawke's Bay wind farm near Napier—75 turbines, 30.0 km², Project West Wind near Wellington—62 turbines, 55.8 km², Project Hayes near the Lammermoor Range, Otago—176 turbines, 92 km². It was found that direct habitat loss from wind farm construction was usually small-scale and unlikely to have a significant impact on bird populations. However, a considerable proportion of habitat may be lost if a particularly scarce and important habitat type was affected, or if there was potential for the effects to extend into the wider area (e.g., through disrupting the hydrology of a wetland).

18.2   Possible bird and wind turbine interactions

18.2.1   Collision fatalities

Direct mortality at wind farms results from birds striking revolving blades, towers, nacelles and associated powerlines and meteorological masts. There is also evidence of birds being violently forced to the ground by turbulence behind the turbine created by the moving blades.

Two wind farm areas have become synonymous with collision fatalities: Altamont Pass in California and Tarifa in southern Spain. Large numbers of raptors have collided with turbines at these sites, including substantial numbers of golden eagles (*Aquila chrysaetos*) at Altamont and griffon vultures (*Gyps fulvus*) at Tarifa, both of which are long-lived species with low reproductive outputs. While the numbers of collisions per turbine at Altamont and Tarifa have been relatively low (considerably less than 1 bird per turbine per year for each), the total number of collisions has been significant, as a result of the large number of turbines. Also and of particular importance, both sites support important food resources that attract raptors, resulting in birds of these species for ageing within the collision-risk zone of turbines. Thus, in both areas, the scale and siting of the wind farms are inappropriate given the species' behaviour (large soaring species with poor flight manoeuvrability), which makes them vulnerable to colliding with turbines and their demographics, which make their populations vulnerable to small increases in mortality.

Most other studies completed to date suggest low numbers of bird fatalities at wind farms. In comparison, studies of bird collisions at coastal wind farms have generally reported higher numbers of collisions, which may reflect higher bird densities at coastal sites, or greater frequency of bird movements at such sites.

Unfortunately, in many instances these numbers are likely to be underestimates, as they are often based only on found corpses, without accounting for scavenging and searcher efficiency. Several studies have indicated rapid removal of carcasses by scavengers. For example, in the U.S., Kerlinger and others found that most passerine carcasses disappeared within 3 days, but that large carcasses remained for at least 1–2 months.

As wind power becomes more popular and wind farms become more abundant, collision numbers will increase. Indeed, given current documented average mortality rates of about 2 bird deaths/turbine/year, the projected impact of turbines in the U.S. could be in the range of 1–5 million birds per year by 2025, if large numbers of wind turbines become part of the landscape. This makes proper siting imperative to help reduce bird mortality and therefore population effects.

An important issue is whether or not the collision fatalities at wind farms are sufficiently great in number to cause population declines. Even when collision rates per turbine are low, collision mortality at a wind farm may be considered high, especially when composed of hundreds or thousands of turbines. The cumulative mortality from multiple wind farms may also contribute to population declines in susceptible species, such as soaring raptors. Furthermore, even relatively small increases in mortality rates may have a significant impact on some populations of birds, such as a threatened species, or a long-lived species with low annual productivity and slow maturity, such as many New Zealand waders, particularly when adults are killed.

The strongest evidence of collision mortality affecting populations comes from studies of particularly vulnerable species that are present in relatively high numbers in the vicinity of wind turbines. The most vulnerable species appear to be those highly susceptible to collision and with low productivity (e.g., large raptors, seabirds), making them less able to compensate for increased levels of adult mortality. For example, a long-term study of golden eagles at Altamont Pass, California, showed that the incidence of collision mortality had reduced productivity in the local population to the point were it had become a sink, dependent on immigration for its maintenance.

Similarly, evidence from a study of nesting terns at Zeebrugge, Belgium, estimated additional mortality of at least 1.5% for two species as a result of colliding with turbines as they returned to their nests. Dierschke and others suggested that such increases in mortality of greater than 0.5% could have serious population impacts.

18.2.2 Factors contributing to avian mortality in wind farm

There appear to be four main (and often interacting) factors that contribute to avian mortality at a particular wind farm site:

1. Density of birds: In general, there are more opportunities for birds to collide with turbines when there is an abundance of birds or high frequency of movements. This does not mean that high bird density or frequency of movements necessarily translates into greater bird mortality, a direct relationship between the number of birds in an area and collision rate has only been documented by one study.

2. Bird species: Particular species or groups of birds appear to be particularly prone to collision with structures such as wind turbines. These groups include swans and ducks (*Anseriformes*), raptors (*Accipitridae*), particularly large soaring species, owls (*Strigiformes*) and nocturnally migrating passerines.

3. Landscape features: Some landforms at wind farm sites, such as ridges, steep slopes, saddles and valleys, may increase the degree of interaction between turbines and birds using or moving through an area, although some debate exists around this point. The presence of other landforms, such as peninsulas and shorelines, can funnel diurnal bird movement, which may also affect collision rates, although this has yet to be studied. These features can combine with high bird abundance to create high collision risk.

4. Poor weather conditions: At many sites, collisions by nocturnal migrants tend to occur during episodes of poor weather with low visibility. Although most examples appear to be isolated incidents, weather conditions should be kept in mind if a wind farm is being proposed in an area that has a large number of poor visibility days (< 200 m visibility) during spring and autumn (periods of migration) and has other confounding factors (e.g., large numbers of nocturnal migrants and landform features such as ridges present).

18.2.3 Habitat loss

Wind farm development will result in habitat loss for birds. Land will be taken up by turbine bases and access roads and secondary effects, such as altered hydrology, are possible. In the U.K., habitat loss or damage as a result of wind farm infrastructure is not generally perceived to be a major concern for birds outside designated sites of national and international importance for biodiversity. Typically, actual habitat loss only amounts to 2–5% of the total development area and careful positioning of turbine bases and routing of access roads, together with the use of proven restoration techniques, should ensure that any loss is minimised. However, the cumulative loss of or damage to sensitive habitats may be significant, especially if multiple large developments are sited at locations of high bird use. Furthermore, direct habitat loss may be additive to displacement.

The scale of habitat loss, together with the availability and quality of other suitable habitats that can accommodate displaced birds and the conservation status of those birds, will determine whether or not there is an adverse impact on populations. The possibility that wintering birds might habituate to wind farm structures has been suggested, but there is little evidence and few studies of long enough duration to show this.

Differences in behaviour between residents and migrants have been observed in some studies, but not in others. Unfortunately, very few conclusive studies are available because most lack well-designed procedures incorporating observations both before and after construction.

18.2.4   Disturbance and displacement

Disturbance and displacement may arise from increased activity by people at a wind farm during construction and maintenance, as well as from improved road access as a result of the wind farm development, especially in areas where there was little human activity before the wind farm existed. Roads may also improve access for predators of ground-dwelling or ground-nesting birds, such as wandering dogs (*Canis lupus*), possums (*Trichosurus vulpecula*) and hedgehogs (*Erinaceus europaeus*). The presence and noise of turbines may deter birds from using an area close to these.

Some studies appear to show little or no behavioural impact of wind turbines on various bird species. In some cases, this apparent lack of evidence may be an artefact of such things as the type and intensity of monitoring. However, in Britain the majority of recent studies have also found no disturbance effects and there is an increasing body of evidence that wind farms generally do not affect bird distribution.

In one studies, a reduction in bird numbers has been reported as far as 600 m from turbines outside the breeding season and up to 300 m from turbines during the breeding season. Such variation was found during two studies on the barnacle goose (*Branta leucopsis*) population. The first study, which was carried out on the birds' spring staging grounds in Sweden, where they fed in close proximity to wind turbines (to within 25 m), found no significant disturbance effect. However, the second study of the same population on their wintering grounds in Germany found that few geese fed within 350 m of turbines and there was a reduction in numbers up to 600 m from the turbines. The most likely explanation for such different results is that geese avoid turbines when there is easy access to alternative feeding habitat, but will be less selective when resources are limited. Similar results of birds becoming more tolerant of disturbance as resources become scarcer have been found in other studies of disturbance of wintering waterfowl and studies to date have shown that substantial displacement by wind turbines seems to have occurred primarily in farmland habitats, where there would typically be alternative feeding areas within easy reach. Other results suggest that disturbance can lead to reduced breeding productivity, reduced survival or a reduction in available habitat, so disturbance may be significant for some species in certain situations.

Studies of birds' responses to turbines at night, using thermal and passive imaging equipment plus radar, revealed that more flight reactions occurred with headwinds (87%) than with tailwinds (29%). Winkelman's observations in daylight indicated that over 75% of all reactions took place within 100 m of the turbines, with ducks reacting at the greatest distance and passerines reacting closest to wind turbines.

Relatively long lines of turbines or large wind farms can become important barriers to the local or seasonal movements of birds. The effect of birds altering their local flight paths or migration routes to avoid a wind farm is a form of displacement. This effect is of concern because it may result in increased energy expenditure when birds have to fly further to avoid a large array of turbines and it may disrupt linkages between distant feeding, roosting, moulting and breeding areas. The magnitude of the effect will depend on species, type of bird movement, flight height, distance between rows of turbines, layout and operational status of turbines, time of day and wind force and direction. The impact can range from a slight 'check' in flight direction, height or speed, through to significant diversions that may reduce the numbers of birds using areas beyond the wind farm.

Several studies have shown that some species alter their route to avoid flying through wind farms, e.g., tufted duck (*Aythya fuligula*) and common pochard (*Aythya ferina*) at Lely in The Netherlands. While this may reduce collision risk, it could result in the wind farm acting as a barrier to bird movements. However, such effects are not universal, for example, at Zeebrugge, large numbers of birds regularly fly through a wind farm without diverting around it and van der Bergh and Everaert and Stienen concluded that a line of turbines did not act as a barrier to the daily flight paths of breeding gulls and terns. In contrast, studies of bird movements in response to offshore developments have recorded waterfowl taking avoidance action between 100 m and 3000 m from turbines. These findings highlight the species- and site-specific nature of wind farm impacts on birds.

Some birds will fly between turbine rows, as seen with common eider (*Somateria mollissima*) at Nysted, where the turbines were 480 m apart. However, their ability to do so will depend on the distance between turbines.

Although evidence for this type of response is limited, these observations have implications for wind farm design. Generally, spacing between turbines at onshore wind farms is recommended to be a minimum of 200 m apart to avoid inhibiting bird movements. This recommended distance is often the minimum spacing required by industry to reduce wake effects of large turbines on neighbouring turbines.

For a small wind farm (< 10 turbines), the ecological consequences of any barrier are unlikely to be a problem, with minimal diversion distances involved. For larger sites, however, the barrier effect has the potential to be more important. Thus, it is important to consider new wind farm proposals on a case-by-case basis and to assess the patterns of resource availability and the potential loss through disturbance for each. However, it should be noted that a review of the literature suggests that none of the barrier effects identified so far have had significant impacts on populations.

18.3 Observed impacts of wind farms on various groups of birds

18.3.1 Habitat groupings

The following is a review of the impacts of wind farms on various groups of birds, largely in relation to the main habitat type they occupy.

Waterbirds

Waterbirds include species that are typical of terrestrial wetland habitats, including ponds, lakes and rivers. This category excludes seabirds, waterfowl and shorebirds, which are discussed separately.

There have been few reports of waterbird fatalities resulting from collision impacts at wind farms, but in many cases the methods used to detect them have been imprecise. Gulls and terns have been identified as being especially vulnerable to mortality due to wind turbines because they often fly within the height of the rotor sweep zone. However, despite their perceived vulnerability, very low numbers of gulls and terns have been reported as colliding with turbines, with the exception of three sites in Belgium. At one of these sites, Zeebrugge, Everaert and Stienen calculated that the mean number of collision fatalities (mainly gulls and terns) per turbine per year in 2004 and 2005 was 20.9 and 19.1 birds, respectively, after taking into account the number of dead birds found under turbines and the correction factors for available search area, search efficiency and scavenging.

The black shag (*Phalacrocorax carbo*) and cattle egret (*Bubulcus ibis*) are the only species of waterbirds occurring in New Zealand that were listed by Kingsley and Whittam as having been found fatally injured after colliding with a wind turbine. However, Kingsley and Whittam did list representatives from several genera such as *Larus* (gulls), *Sterna* (terns), *Ardea* (herons) and *Nycticorax* (night heron). Three such waterbird species occasionally forage over pasture near wetlands and are threatened, the red-billed gull (*Larus novaehollandiae*) (gradual decline), black-billed gull (*Larus bulleri*) (serious decline) and black-fronted tern (*Sterna albostriata*) (nationally endangered). Therefore, any wind farms sited in pastureland that may have deleterious impacts on the populations of these three species would be of concern.

Seabirds (order Procellariiformes)

Procellariiformes, particularly the larger species, may be just as vulnerable to turbine collision fatalities as soaring raptors, because these seabirds are adapted to sustained high-speed flight with slow manoeuvrability in unobstructed environments. In addition, many have delayed maturity and low productivity, making their populations sensitive to increased mortality.

Two species fly some distance inland to their colonies the nationally endangered Hutton's shearwater (*Puffinus huttoni*), which flies to the Seaward Kaikoura Range and the range restricted Westland Petrel (*Procellaria westlandica*), which flies to the coastal foothills of the Paparoa Range. Obviously, any turbines erected in the flight paths of these two species, both of which have restricted colony distributions, would be highly likely to result in collision fatalities. In addition, both species fly to and from their colonies at night, particularly around dusk and dawn. It has been found that nocturnal seabirds, especially fledglings, can become disorientated, especially during periods of fog and are then prone to being attracted to artificial lights, such as street lights. Thus, lighting on turbines would increase the risk of collision for these nocturnally active seabirds if wind farms were sited near their colonies or on routes between the sea and their colonies.

Waterfowl

The effects of wind turbines on waterfowl (e.g., ducks, shelducks, geese and swans) have been examined at a few wind farms, particularly in Europe. Even though waterfowl are regarded as prone to collision with turbines, the presence of large numbers of waterfowl near wind farms does not necessarily mean that large numbers of fatalities will eventuate. In some cases, seaducks are believed to have learned to avoid turbines, resulting in fewer collisions over time. Sites in the U.S. with year-round waterfowl use reported the most fatalities of dabbling ducks (*Anatinae*) and at these sites waterfowl made up 10–20% of all fatalities. However, numbers of fatalities were still low, especially in relation to the number of ducks that used the areas. Disturbance is an important factor to consider when siting a wind farm near significant waterfowl areas.

The most comprehensive study of the effect of wind turbines on waterfowl took place in Denmark and involved a modern, 10-turbine offshore facility in an area where large numbers of common eider (*Somateria mollissima*) and black scoter (*Melanitta nigra*) fed. It was found that these diving ducks exhibited avoidance behaviour towards the turbines, which was accentuated in poor weather. Eiders generally avoided flying or landing within 100 m of the turbines and avoided flying between turbines that were spaced less than 200 m apart, preferring to fly around the outer turbines. Similarly, two diving duck species, common pochard and tufted duck, were tracked at night using radar and were found to avoid flying near turbines, passing around the outer turbines instead. In a meta-analysis of 19 studies into the effects of wind farms on bird abundance, Stewart and others found that wind farms seemed to reduce the abundance of many bird species and that Anseriformes (swans, geese, ducks) experienced greater declines than other bird groups, suggesting that a precautionary approach should be adopted to wind farm developments near aggregations of Anseriformes.

Shorebirds

In North America, observed mortality of shorebirds (waders) at wind farms has been low, possibly because few sites are located in shorebird habitat. In contrast, Stewart and others found that wind farms can have a negative impact on the abundance of shorebirds and advocated a precautionary approach to wind farm development at coastal sites where aggregations of shorebirds occur. This result was derived from a meta-analysis of six studies two in the U.S. and one each in Germany, The Netherlands, Scotland and England.

Each species of shorebird appears to have a different threshold to disturbance. For example, at Blyth Harbour wind farm in the UK, purple sandpipers (*Calidris maritima*) did not seem to be disturbed by either the construction process or the operation of wind turbines. In contrast, studies in the Netherlands and Denmark examining the effect of turbines near important staging areas for many shorebird species found that the birds avoided the turbines and were at a relatively low risk of collision. Some studies have shown that shorebirds avoid turbines up to 500 m away, while others have shown no significant effect on shorebird distribution. It is not known whether this inconsistency in behaviour between species is related to the abundance and proximity of alternative suitable habitat, a species may be more likely to move away from turbines if there is ample suitable habitat nearby.

Diurnal raptors

Collision has been the focus of raptor studies at wind farms, due to the high collision rates observed at a small number of sites. One study at Altamont,

California, U.S., which involved observations and carcass searches over six seasons and covered 16% of the 7000 turbines, found 183 dead birds (0.05 birds per turbine per year), 65% of which were raptors. Of these deaths, 55% were attributed to turbine collisions, 8% to electrocution and 11% to wire collision, for 26%, the cause of death could not be determined. There has also been significant raptor mortality at Tarifa, Spain (0.34 birds per turbine per year). This site is near the Strait of Gibraltar and forms a bottleneck that concentrates bird migration between Europe and Africa in the Mediterranean basin, at least 30000 raptors and large numbers of storks pass through the area each autumn.

Very few raptor fatalities have been reported at other sites. In parts of the U.S. outside California, raptors comprised only 2.7% of turbine-related deaths. However, even though this percentage seems small, an increase in mortality of greater than 0.5% could have a serious impact on a population of long-lived raptors with low productivity.

The most important factor that influences raptor collision rate appears to be topography, in particular elevation and the presence of ridges and slopes. The low numbers of raptor fatalities observed at the majority of wind farms is most likely due to improved siting of turbines, away from problem topography and high raptor concentrations. It has been speculated that the construction of tubular (as opposed to the lattice type) towers and slower rotor speeds may also have helped to lower raptor fatalities, but no studies to date have shown a significant relationship between mortality levels and turbine type. It was found that the high mortality at Altamont and Tarifa resulted from a combination of sensitive species (soaring raptors) flying through the area in large numbers (important feeding areas and migration route, respectively) and turbine layout (hundreds in densely packed formation) and design (lattice towers attractive to raptors as perches).

Landbirds

Amongst the landbirds, passerines are the group most commonly affected by wind farms in parts of North America outside California. Protected passerines comprise 78% of all fatalities documented at wind farms in the U.S. This proportion would be even greater if it included unprotected species, such as the starling (*Sturnus vulgaris*) and house sparrow (*Passer domesticus*). Grassland bird species with aerial courtship displays, such as the horned lark (*Eremophila alpestris*), appear to be particularly prone to collisions with turbines, as they fly high enough when displaying to collide with turbines. However, during migration most passerines fly at night and at an altitude in good weather (1000–1500 m) that takes them well above turbine height.

The greatest threat from wind farms to migrant passerines in North America was found to be habitat loss. In contrast, the impact of turbines on forest-nesting passerines was found to be low, with several nesting in the forest within 20–30 m of the turbines, although a few species were found to avoid clearings where turbines were located and some appeared to move further into the forest. However, since there has only been one study to date into the effect of wind turbines on forest-nesting birds, more studies are needed to understand these effects. Turbines may displace some grassland species of landbirds. Leddy and others found that there were fewer nesting grassland birds within 100–200 m of turbines than beyond and densities decreased by more than 50% within 50 m of turbines. In contrast, Devereux and others found that the distribution of four functional groups of wintering farmland birds (granivores, corvids, gamebirds and the skylark *Alauda arvensis*) was unaffected by turbines in East Anglia, England (in 150-m-wide blocks), at distances ranging from 0 m to 750 m. They also measured occurrence in areas 0–75 m and 75–150 m from the turbines and found no evidence that the four functional groups of farmland birds avoided areas close to turbines.

Gamebirds (pheasants and quail in New Zealand), which are a subset of the landbirds group, are vulnerable to habitat destruction and fragmentation and disturbance of local breeding populations as a result of human-induced changes in the landscape, such as wind farm developments. In North America, much of the remaining suitable habitat for gamebird species is located in remote areas or where topography makes agriculture difficult.

Some of these sites may be suitable for wind farms and so turbines and associated structures could adversely affect sensitive and vulnerable gamebird species. In agreement with this conclusion is the finding of Devereux and others that the distribution of the pheasant (*Phasianus colchicus*) was negatively effected by turbines. S.M. Percival considered that there is a low risk of gamebirds colliding with turbine towers.

The feral pigeon (*Columba livia*), rook (*Corvus frugilegus*), skylark (*Alauda arvensis*), blackbird (*Turdus merula*), song thrush (*Turdus philomelos*), starling, chaffinch (*Fringilla coelebs*), greenfinch (*Carduelis chloris*) and house sparrow are landbird species that occur in New Zealand and were listed by Kingsley and Whittam as having been found fatally injured after colliding with wind turbines. In addition, the genera *Hirundo* and *Anthus* are represented in their mortality list, both of which have representative species in New Zealand (welcome swallow *H. tahitica* and New Zealand pipit *A. novaeseelandiae*). Most species mentioned above are introduced and none are threatened.

The California quail (*Callipepla californica*), chukor (*Alectoris chukar*) and pheasant (*Phasianus colchicus*) are gamebird species that occur in New

Zealand and were listed by Kingsley and Whittam as having been found fatally injured after collision with wind turbines. All of these gamebirds were introduced to New Zealand and all except the chukor are widely distributed.

18.3.2 Seasonal groups

Breeding birds

In general, birds breeding near wind turbines have been reported to have lower collision rates than non-residents. In part, this is probably because local birds become familiar with turbines, whereas individuals passing through the area would not have that familiarity and may be unable to detect turbines before a collision occurs if weather conditions are poor, e.g., during fog. However, wind farms are likely to have a greater impact on breeding birds as a result of habitat loss, obstruction of regular flight paths, disturbance by people servicing turbines and obstruction to important feeding areas (particularly important in coastal areas). Bird productivity (breeding success) does not appear to be negatively affected at many wind farms.

For example, in one study, mean productivity at a 66-turbine site, was the same as in surrounding areas. However, few such studies have been carried out.

Reduced breeding bird populations were noted at a few wind farms where breeding habitat was destroyed during installation of turbines and where people and vehicles were continuously present in the area. It has also been found that many grassland birds avoid nesting within 100–200 m of turbines. Ketzenberg and others investigated the breeding densities and spatial distribution of the common skylark (*Alauda arvensis*) and some species of breeding waders (Eurasian oystercatcher *Haematopus ostralegus*, northern lapwing *Vanellus vanellus*, common redshank *Tringa totanus* and black-tailed godwit *Limosa limosa*) before and after installation of wind farms in four coastal areas in Lower Saxony, Germany.

They found no consistent pattern in the change in number of breeding pairs following construction, with some decreases but also some increases for some species of waders, the numbers increased near wind turbines because of the change in farming practice post-construction, emphasising the need to consider other changes contemporary with wind farm development. Similarly, there was no significant difference in numbers of breeding pairs of ducks (*Anatinae*), waders (*Charadriiformes*), Arctic skua (*Stercorarius parasiticus*), gulls (*Laridae*) and small passerines between the year of installation of a 3-turbine cluster and the subsequent 8 years at Burgar Hill, Orkney Islands.

Many seabirds, including coastal species such as gulls and terns, are readily disturbed by the activities of people near their breeding colonies, so that the presence of turbines may cause the abandonment of a site.

Wintering birds

The numbers and movements of sedentary species remain much the same year round, particularly for most forest-dwelling and open-country species. However, physical or biological factors, such as localised habitat and/or food supplies, may act to concentrate birds such as waterfowl and shorebirds. Thus, depending on the site of a wind farm, bird densities in the vicinity may remain much the same, increase or decrease during winter. For example, studies at Netherlands, found reductions in density within a wind farm area in winter for four duck species (mallard *Anas platyrhynchos*, tufted duck *Aythya fuligula*, common pochard *A. farina* and common goldeneye *Bucephala clangula*), which extended to 300 m away from the farm. In contrast, there was little or no effect on great-crested grebe (*Podiceps cristatus*), Eurasian coot (*Fulica atra*) or common gull (*Larus canus*) and increased numbers of black-headed gulls (*Larus ridibundus*) and greater scaup (*Aythya marila*). At Blyth Harbour wind farm, U.K., great cormorants (*Phalacrocorax carbo*) were temporarily displaced from their roost during construction, but returned once the farm was operational. Numbers of great cormorants, common eiders (*Somateria mollissima*), purple sandpipers and gulls were comparable before and after construction. This wind farm is sited in a commercial harbour and comprises nine turbines built at 200-m intervals along the estuary's breakwater. The harbour is a site of special scientific interest because it hosts a large winter roost of the purple sandpiper and the estuary it protects adjoins a Ramsar site.

Wind farm layout can also affect avoidance behaviour. For example, for pinkfooted geese, the avoidance distance was 100 m for lines of turbines, compared with 200 m for clusters of turbines and geese did not enter the area between turbines arranged in a cluster.

Migrating birds

Although long-distance movements of birds can occur in any month, the periods of peak migration in New Zealand occur in spring, summer and autumn. Different species and possibly different age and sex categories of the same species, migrate through the same area during different periods. Migration can also occur in winter, e.g., northward movements following unusually severe southerly storms that bring snow to sea level. In summer, there can also be movements of subadult birds or failed breeders from nesting areas to staging areas (coastal sites), or to wintering sites further north. Thus, the pattern and timing of migration can be highly unpredictable. The broader the spatial and temporal scale, the more predictable migration movements appear, but with regard to a particular local area on a given day, it is very difficult to predict whether migrants will be present.

Meteorological conditions can have a large influence on the numbers of birds involved in migration. In Canada, numbers of birds migrating have been shown to vary 10-fold or even 100-fold from one day or night to the next, depending largely on weather. A bird may migrate several hundred kilometres in a day or night when the weather is favourable and then may not migrate for several days when the weather is poor. Migrant numbers appear to be greater at times with (or following) light tail winds than when winds are strongly opposing. Such winds allow birds to travel a given distance more quickly and with less energy expenditure than would be required while flying into a headwind. There is also a close interaction between migration and other weather variables such as temperature, humidity and pressure and it is not well established which specific variables cue birds to migrate rather than remain on the ground. In the case of migrants, flights once underway tend to be at high altitude, well above turbine height, to maximise flight and energy efficiency. Birds wait for suitable conditions before embarking on migration, but may be forced to lower their flight altitude if they encounter bad weather during migration. Therefore, migrants are at risk of collision with wind farms mainly during takeoff and descent, when their flight paths take them through the height range of the rotor-sweep zone.

Many collisions reported at wind farms in North America involve migrating birds. For example, Johnson and others noted that 71% of carcasses were migrants. Sites in different regions differ in the magnitude of bird migration and the influences on this migration. For example, in western North America, there is little evidence that tall human-made structures kill large numbers of night-migrating birds, whereas this is a well-documented phenomenon in eastern North America.

The reason for this regional difference is unclear, although it may be due to lower densities of nocturnal migrants in the west, or differing meteorological conditions leading to different avian behaviour. Whatever the reason, this is an important point that must be considered when comparing mortality studies from sites outside the general area of a proposed wind farm.

Inclement weather can increase the risk of migrant collision with wind farm structures. For example, a cloud ceiling that drops to near or below the height of turbines will affect high-altitude migration, inducing migrants to move at or below treetop level and therefore increasing the probability of collisions with tall obstacles. Drizzle and fog impair visibility and cause birds to fly at lower altitudes and follow topographical cues. The combination of such weather with lighting at wind farms may attract migrating birds and so increase the collision rate. Thus, if there is a high proportion of foggy days during a period of migration at a proposed wind farm site that is on a migration route, there is likely to be an increased risk of collision.

Wind farms situated on prominent landforms can also represent greater potential risks to migrating birds. Features that rise abruptly in the landscape, such as high ridges and mountains, can influence bird movements and if wind farms are sited at high elevations, turbines may end up at a height that enters the altitudinal strata typically used by migrants. For example, the turbine rotor sweep zone of 100 m towers located on a ridge 200 m above the surrounding landscape are effectively 300 m in the air and at an altitude where nocturnal migrants may be flying.

Diurnal migrants

Some groups of birds, e.g., raptors, are principally diurnal migrants. Diurnal migrants that use thermals (rising warm air caused by the sun heating the earth) to reach their preferred altitude do so to facilitate soaring and conserve energy. As a result, the number of such migrants tends to decline in the late morning and through the afternoon. Diurnal migrants can be more constrained by topographical features than nocturnal migrants and tend to concentrate along linear features, such as coastlines, rivers, ridges and valleys. Birds will often divert by as much as 45° from their preferred course in order to fly along such a 'leading line'. The greatest concentration of birds often occurs at these features when there is a crosswind relative to that feature. Therefore, the placement of wind farms on such topographical features may result in interactions with diurnal migrants.

Nocturnal migrants

Many bird species migrate at night (e.g., grebes, ducks, rails, waders, cuckoos). There are three main reasons why birds flying at night collide with wind turbines and these are often inter-related height of the structure (and the landform it is located on), lighting and weather.

Many U.K. and North American nocturnal migrants continue to migrate for at least part of the day, but do so at lower altitudes, tending to stay within 20–30 m of the ground (within or near vegetation) to avoid predation. On a typical day during migration, birds move between higher and lower altitudes at dawn and dusk and it is during these times that birds may be at risk of colliding with wind farm structures. At daybreak, or just before it, nocturnal migrants drop rapidly from higher altitudes (> 200 m) and fly at or above treetop level (< 200 m) until they find a suitable location for landing, features of which will depend on the conditions and the requirements of the individual birds. There appears to have been only one comprehensive study calculating the collision risk for nocturnal migrant birds. This was performed in the Netherlands and collision risk was calculated by means of observed collisions (using thermal image intensifiers). The results showed a high nocturnal collision probability, with 1 in 40 (2.5%) birds passing at rotor height. Daily searches for collision fatalities during the migration periods, together with systematic field observations of passing birds, could lead to a better picture of the behaviour and collision risk of birds. The use of night vision devices and/or radar and thermal image intensifiers are regarded as necessities.

Staging areas

Some types of migrants, such as shorebirds and waterfowl, flock at restricted areas of suitable habitat while resting and feeding between migratory flights. These 'staging areas' are often lakes, marshes, estuaries, mud flats or other areas that can provide food and/or shelter for large numbers of birds. Once a migrant decides to stop, it is constrained by the availability of habitat and resources within the local landscape. Stopover sites are not necessarily large expanses of high-quality habitat, such as mudflats where thousands or millions of birds congregate, they can also include marginal habitat when nothing else is available in the immediate area. For example, a flock may be forced to land and stopover at a marginal site during bad weather. At staging areas, flights of migrants are often concentrated into corridors when the birds are either taking off or approaching to land. The flight height of these migrants is often at the height of wind turbines. Some birds, like swans, typically climb only very gradually and may remain low for a considerable distance after takeoff from the stopover area, while other birds climb more rapidly. Therefore, the distance from the stopover area within which flight altitudes will be low enough to be at risk of collisions with turbines will depend on the species. Collision with wind farm structures is not the only potential effect on migrating birds. Disturbance can also affect migrants if turbines are located near important staging areas. Additionally, the alteration or destruction of habitat used by birds during migration can also contribute to adverse environmental effects.

18.4   Mitigation of impacts

The most useful way to ensure minimal negative effects of wind farms on birds is to choose an appropriate site. However, a number of mitigation measures have been suggested to reduce collision fatalities at operational wind farms, although it must be emphasised that most have yet to be tested to determine their effectiveness. Mitigation may involve on-site and/or off-site measures. Temporary shutdowns of turbines during periods of high bird activity, especially at migration bottlenecks and staging areas and near breeding or wintering concentrations, have been proposed. Since turbine shutdown has yet to be routinely implemented, it is not known to what extent it would reduce collision fatalities, although stationary blades are likely to pose less of a risk to flying birds than rotating blades. However, because collisions also occur with turbine towers, this does not remove the need to avoid siting wind farms on migration routes or at other sites where concentrations of species vulnerable to collisions occur. In this regard, it is of note that in response to a 2004 lawsuit filed against the Altamont turbine operators over raptor kills, wind-power companies and local county officials agreed to shut down half the turbines during winter months and permanently remove 100 turbines over 5 years.

It has been suggested that scaring devices, such as playback of alarm calls, could be used as a deterrent. However, this is likely to be of short-term effectiveness and unacceptably intrusive close to human habitation. Radar- or audio-activation of possible risk-reduction measures, such as alarm calls or turbine shutdown, has the potential advantage that it could be initiated when a hazardous situation is developing, as birds approach. However, given that such scaring devices have not been trialled at wind farms, much development and testing would be required before they could be accepted as an effective method for deterring bird species from wind farms in New Zealand.

It has been proposed that the visibility of rotating blades to birds could be increased by having high contrast patterns on blades. This proposal requires field testing, but even if it reduced collision risk, such obvious turbine blades visible from urban areas may not be acceptable to the general populous. The use of ultraviolet paint has also been suggested as potentially helpful in alerting birds to the presence of rotors while not increasing their visibility to people. However, results from limited trials have been equivocal, perhaps because of different species' sensitivities to different UV wavelengths. Smith and others found that turbines at the ends of lines and edges of clusters killed disproportionately more birds and so hypothesised that a pair of poles could serve as dummy turbines beyond the end of lines and edges of clusters. These poles would be placed 5–10 m apart, just beyond the rotor plane of the end turbine and upward to the maximum height of the rotor. These 'flight diverters' would be expected to encourage birds to fly around or over the operating turbines. Another suggestion to overcome this problem is to relocate turbines that kill disproportionately more birds because of where they are located.

Another suggested mitigation measure could involve adjusting turbine tower height to minimise collision rates. Taller or shorter towers could expose fewer birds to collision, although little research has been conducted on this factor. It would require detailed knowledge of the variability of flight altitude of species prone to collision mortality at the site to determine whether such an adjustment would be effective.

Reducing collision mortality of resident species could involve making the site unsuitable for use by birds or a specific bird species through changes in habitat. This action has been effective in reducing bird abundance on grassed airfields, where mown swards were made unsuitable to foraging and roosting species by being left to grow long (>230 mm). Off-site mitigation can involve actions taken to increase the security of at-risk species at sites away from wind farms. This might involve creating or improving habitat near a wind farm to encourage birds to use it rather than the wind farm site.

An alternative procedure could involve management to improve adult survival or fledgling production, e.g., by carrying out mammalian predator control for New Zealand species. Ideally, where an assessment has quantified the level of adverse effect on a bird population, there may be an opportunity to carry out management to mitigate against such effects. An essential aspect of any mitigation measure would be to monitor its impact and test its effectiveness in either reducing collision fatalities or increasing numbers of individuals above those lost to collision fatalities.

18.5   Summary and conclusions

The effects of wind farms on birds are variable and can be species-, season- and site-specific.

1. The four main factors that contribute to collision fatalities at a wind farm are high densities of birds or frequency of movements through it, presence of species prone to collision with turbines, landscape features that concentrate bird movement and poor weather conditions.

2. Species groups that are most prone to collision fatalities at wind farms in Europe and North America are herons and allies, swans, geese, ducks, large soaring raptors, gulls, terns, owls and nocturnal migrant passerines.

3. While carcass numbers found at wind farms have been documented, these will underestimate fatalities unless a systematic methodology is used, including taking into account scavenger rate and searcher efficiency.

4. Loss of or damage to habitat as a result of wind farm construction (roads, turbines, buildings) tends to be a minor impact, unless sensitive or rare habitats are involved, or habitat management at the site changes as a result of the development.

5. Disturbance of birds as a result of wind farm development and operation may arise from increased activity of people and/or the presence, motion or noise of turbines. Disturbance may lead to displacement or exclusion of birds from areas of suitable habitat. The degree of disturbance can be highly variable, depending on the bird species, wind farm layout and availability of alternative habitat nearby.

6. The choice of an appropriate site for a wind farm is the most useful way to ensure minimal negative effects on birds.

7. The amount and extent of ecological baseline data collected at a proposed wind farm site should be determined on a case-by-case basis. A minimum of 3 years of detailed investigation should be carried out to determine which bird species use the site and how and when they use the site.

8. Any detailed study should ensure that seasonal, annual and weather variables are suitably investigated, particularly if a site is found to be used by a species that is threatened or likely to be at risk of disturbance or collision by an operational wind farm.

9. Wind farm layout is probably important in reducing disturbance and collision risk to birds. It has been suggested that wide corridors between clusters of closely spaced turbines is the most appropriate layout to minimise collision fatalities and prevent barrier effects for both resident and migrant birds. However, a line formation parallel to the main flight direction of migrants has also been suggested.

10. Wind farm developments should ensure that blade revolutions per minute are as low as possible, to avoid motion smear and thus promote blade visibility during the day.

11. Bright white lighting is regarded as the main attractant of nocturnally active birds leading to collision with tall buildings, so its use should be avoided at wind farms. Ideally, the intensity of lighting should be minimal and be white and flashing, with the interval between flashes being as long as possible. In New Zealand, the lighting required on turbines is specified by the civil aviation authority on a case-by-case basis.

12. Although a number of on-site mitigation measures have been suggested to reduce collision fatalities at operational wind farms (e.g., temporary shutdown of turbines, bird scaring devices, high contrast patterns or UV paint on blades, flight-diverter poles and adjustments to tower height), almost all have yet to be tested in the field to determine their effectiveness, therefore, these should be considered with caution. Off-site mitigation measures could involve habitat management to encourage birds to use sites away from wind farms and/or to improve adult survival or fledgling production.

13. Post-construction monitoring at New Zealand wind farms has been inadequate to accurately determine bird fatalities as a result of collision with turbines because neither systematic search procedures nor trained staff have been used. Fatalities have been reported to involve magpies, gulls, blackbirds and a kingfisher, but these results are probably not indicative of the full range of species killed.

14 In addition, each wind farm site tends to be a little different from any other because of variation in topography, weather, habitats, land use and bird species present.

15. Pre-construction assessments with regard to birds should always be carried out, but the complexity of the assessment required will depend on various attributes of the site, such as the bird species present, their threat status, collision risk and vulnerability to disturbance. Post-construction assessments should always be carried out when threatened or vulnerable species are likely to be using the site, or population impacts are likely to occur.

Carbon footprint of wind energy

19.1   Introduction

All electricity generation systems have a 'carbon footprint', that is, at some points during their construction and operation carbon dioxide (CO_2) is emitted. To compare the impacts of these different technologies accurately, the total CO_2 amounts emitted throughout a system's life must be calculated. Emissions can be both direct – arising during operation of the power plant and indirect – arising during other non-operational phases of the life cycle. Fossil fuelled technologies (coal, oil, gas) have the largest carbon footprints, because they burn these fuels during operation. Non-fossil fuel based technologies such as wind, photovoltaics (solar), hydro, biomass, wave/tidal and nuclear are often referred to as 'low carbon' or 'carbon neutral' because they do not emit CO_2 during their operation. However, they are not 'carbon free' forms of generation since CO_2 emissions do arise in other phases of their life cycle such as during extraction, construction, maintenance and decommissioning.

A 'carbon footprint' is the total amount of CO_2 and other greenhouse gases, emitted over the full life cycle of a process or product. It is expressed as grams of CO_2 equivalent per kilowatt hour of generation (gCO$_2$eq/kWh), which accounts for the different global warming effects of other greenhouse gases.

Carbon footprints are calculated using a method called Life Cycle Assessment (LCA) and is also referred to as the 'cradle-to-grave' approach. This method is used to analyse the cumulative environmental impacts of a process or product through all the stages of its life. It takes into account energy inputs and emission outputs throughout the whole production chain from exploration and extraction of raw materials to processing, transport and final use. The LCA method is internationally accredited by ISO 14000 standards. The robustness of the method means that although carbon footprints vary between individual power plants, the ranking of electricity generation technologies does not change with different sources of data.

19.2   Wind energy

Electricity generated from wind energy has one of the lowest carbon footprints. As with other low carbon technologies, nearly all the emissions occur during the manufacturing and construction phases, arising from the production of steel for the tower, concrete for the foundations and epoxy/fibreglass for the rotor blades. These account for 98% of the total life cycle CO_2 emissions. Emissions generated during operation of wind turbines arise from routine maintenance inspection trips. This includes use of lubricants and transport. Onshore wind turbines are accessed by vehicle, while offshore turbines are maintained using boats and helicopters.

The manufacturing process for both onshore and offshore wind plant is very similar, so life cycle assessment shows that there is little difference between the carbon footprint of onshore ($4.64gCO_2eq/kWh$) versus offshore ($5.25gCO_2eq/kWh$) wind generation. The footprint of an offshore turbine is marginally greater because it requires larger foundations.

19.3   Life cycle and carbon emissions of wind power

There is a significant diversity of views on the life cycle levelised costs and carbon emissions of energy technologies, including onshore wind.

Understanding the economics of wind energy is vitally important to ensure a rational discussion about the role of wind power within the energy mix. The challenge is that 'cost' means different things to different people, with often conflicting views apparently supported by 'evidence'. In part, this is due to confusion about current and future costs of generation, what is included or excluded from estimates and the characteristics of wind relative to other generation types. Additionally, there is conflation of 'costs', 'prices' within the power markets and 'subsidies'.

Carbon emissions: Another key issue of debate is the extent to which onshore wind farms achieve a net carbon emissions reduction over their lifetime. The carbon emissions reduction of wind power cannot simply be estimated as equal to the carbon emissions of conventional coal- or gas-fired generation that it displaces – firstly, wind power generation is not zero carbon, as greenhouse gases are emitted during installation, maintenance and decommissioning, secondly, wind power will not replace all forms of conventional generation equally. The true carbon emissions displacement will therefore depend upon a combination of factors including:

1. The types of power generation being replaced.
2. Any decrease in efficiency of conventional plant operating at part load.
3. The impact of any increase in frequency of start-up and shut-down of conventional plant.

There may also be longer-term impacts associated with the installation of new conventional plant to back up an increase in installed wind capacity.

Wind farm life cycle: This study has analysed various estimations of the costs and greenhouse gas emissions associated with on- and offshore wind throughout its life cycle. The lifecycle stages considered are illustrated in Fig. 19.1.

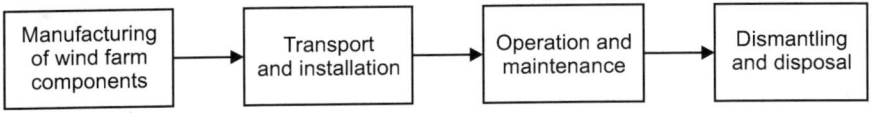

Figure 19.1: Life cycle of a wind farm.

19.4   Life cycle costs

A wide variety of costs are associated with wind farms and these can be grouped into several life cycle components:

1. Capital costs: The fixed costs of construction including manufacturing, installation and transport.

2. Operation and maintenance costs: Annual fixed costs associated with running the farm.

3. Decommissioning: The cost of taking the plant out of commission, dismantling and remediation.

Whilst technologies may be compared on the basis of any of these costs categories, a more holistic view of 'cost' can be gained by looking across the life cycle of the technology and considering their overall cost.

19.4.1   Levelised cost of energy

The Levelised Cost of Energy (LCOE) measures the overall life cycle costs of a technology per unit of electricity produced. It is calculated as the sum of the discounted costs over the generator's lifetime, spread across the discounted units of energy produced over the lifetime. This requires future costs to be expressed in 'present value' terms by discounting.

The calculation of LCOE requires a substantial number of factors to be determined which can be split into those that determine cost and those that determine energy production. Figure 19.2 shows the main information that is required to estimate the costs and energy production of a typical wind farm.

19.4.2   Variation in estimates

There is substantial scope for variation in the estimation of LCOE, which is introduced by the use of different assumptions, methods and uncertainty. These are illustrated in Fig. 19.3 and can be divided into four categories:

1. Variation in input data arising from the scenarios used, timing and locations and uncertainty in the data itself.

2. Uncertainties introduced by the financial assumptions, arising from location, such as tax rates and treatment, prevailing financial treatments, whether pre-or post-tax rates are used and adjustments for risk or inflation.

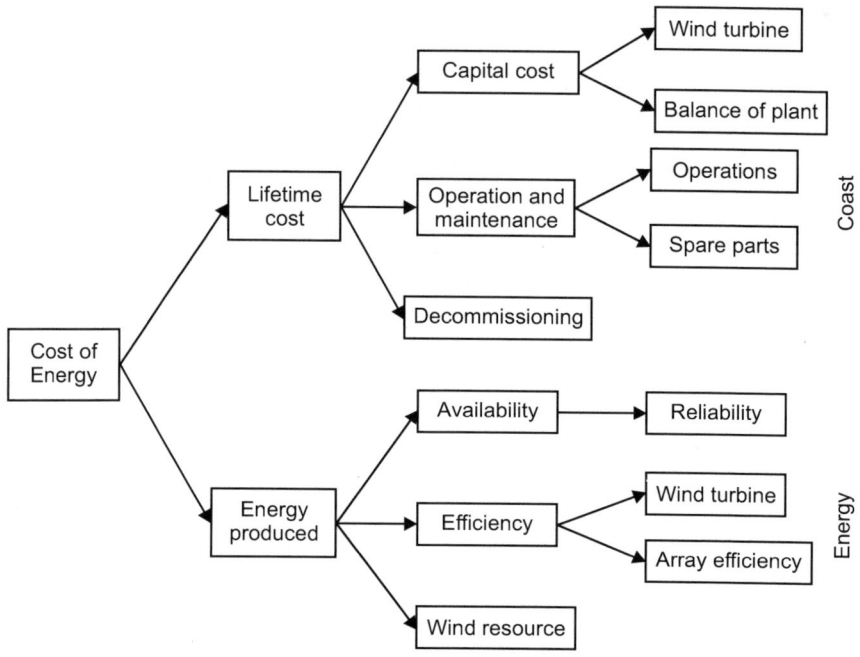

Figure 19.2: Cost of energy for a wind farm.

3. The physical and temporal boundaries analysed and whether specific cost categories are included or not.

4. Differences in the methodology used and its intended scope.

19.4.3 Current cost estimates

In recent years there have been a series of studies providing estimates of costs for wind and comparator technologies. While capital costs do not provide a complete picture, they are the dominant determinant of LCOE for onshore wind, typically accounting for 80–90% of overall life cycle costs. Analysis of key studies showed variation arising from the year of study and location. Most studies offer a range of costs although, as there is limited consistency in terms of how uncertainty in capital costs is reported, interpretation requires care.

Capital costs for offshore wind represent a lower proportion of overall costs (60–80%) with site conditions (e.g., water depth, distance from shore, etc.), having a major impact on costs.

Levelised costs

The variations in capital and operating costs feed through into the overall levelised cost of energy estimates. Here they are joined by a series of other

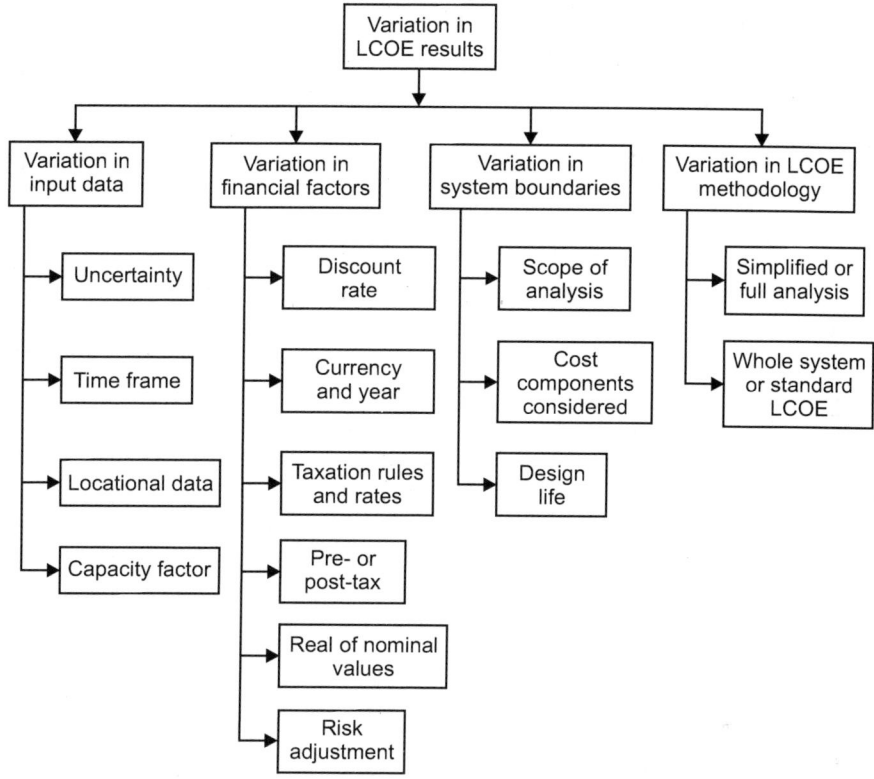

Figure 19.3: Causes of variation in LCOE for wind farms.

factors that lead to significant variation in LCOE. On analysis several things are apparent:

1. Higher values of capital costs do not automatically translate into higher LCOE.

2. U.K. specific studies tend to show higher LCOE values than those for overseas.

3. The spread of costs is large with two studies indicating much higher LCOE for both on- and offshore wind.

The levelised cost of wind is very sensitive to assumptions on capacity factor, lifetime, discount rate and financing structure as well as capital costs. Giberson illustrates this well by making a series of 'reasonable' adjustments (reducing capacity factor, raising discount rates, altering treatment of depreciation). For the studies examined here, the central values for LCOE are strongly correlated with capacity factor and discount rate as well as capital cost.

Comparison with other generating technologies

Many studies offer comparisons between wind and other technologies. It is apparent that there are substantial uncertainties around all technologies. Capital cost, capacity factor and discount rate are important for nuclear, while fossil fuel and carbon costs are important factors for CCGT. It should be noted that LCOE estimates generally assume that thermal power plants are baseload, with capacity factors of 85–90%, in practice not all thermal generators operate as baseload and capacity factors for gas and coal generation would be expected to decline as more wind enters the system, raising their own LCOE.

Although it is evident that offshore wind is substantially more expensive at present, the overlapping of the ranges for nuclear, onshore wind and combined cycle gas turbines means there is no clear outcome in terms of which technology is currently 'cheapest' on the basis of levelised costs. In terms of the evolution of levelised costs over the coming decades, a wide range of studies offer estimates of how capital, operation and fuel costs as well as operational performance will change. The methods, scenarios and assumptions vary, impacting on the projected LCOE. There appears to be scope for reductions in costs of onshore wind arising from advances in turbine technology, manufacturing and turbine performance. Offshore wind has very substantial opportunities for gains from economies of scale, performance improvements, improved deployment and servicing approaches and reduced numbers of offshore cables as a result of a shift to high voltage DC systems. Furthermore, the expectation is that, as deployment increases and practices mature, the risk associated with offshore wind will decrease, driving the discount rate and LCOE downwards.

19.4.4　System costs

The impact of wind on other generators and the system as a whole is generally excluded from levelised cost calculations, although some studies do include them. Some studies that include 'system costs' use it as evidence that wind energy costs are 'significantly understated [because] they failed to take its unusual indirect and infrastructure costs into account'. In essence the 'system' costs that are referred to are:

1. The costs of balancing the power system arising from wind.
2. The costs of providing 'backup' or ensuring there is sufficient generation capacity to meet demand.
3. The cost of additional transmission required to connect wind plants and the associated losses.

Studies show that generation mix, network capacity and interconnection, as well as the availability of mechanisms for managing variability, are important in determining costs, this makes comparison challenging.

Balancing

The variable nature of wind power, in contrast to conventional, dispatchable technologies, requires flexible 'reserves' to be on hand for when the resource is not available. Reserves are used to handle unpredicted variations in demand or generation on a range of timescales from seconds to around four hours. Reserves are provided by power stations running at part load, standby generators that can be started quickly (e.g., open cycle gas turbines) as well as (some) contracted demand response. Costs are incurred by operating power plants less efficiently and ensuring standby generation is available. The amount of reserve is specified by National Grid on the basis of the largest generator that can be lost and the level of error in forecasting demand and wind four hours ahead of delivery. Increases in wind capacity will therefore increase the amount of reserve that needs to be held but the amount depends on overall expected errors, not simply wind capacity.

Overall, the literature suggests that balancing costs are likely to be lower in larger markets, with a geographical spread of plants and when wind is part of a complementary portfolio of other generation technologies. This is important in considering wind integration in Scotland as, while Scotland's wind penetration will be locally very high, it is the penetration at GB level and the extent of transmission and external interconnections that will strongly govern balancing costs. While there are undoubtedly additional balancing costs arising from integrating variable wind, the IEA and other studies suggest they are not prohibitive. Additionally, the Committee on Climate Change (ICC) suggest that, with the right investment in flexibility in the form of storage, demand side management and interconnection, costs can be managed at even relatively high renewable penetration.

Backup

Ensuring there is sufficient generating capacity to provide secure electricity supply is a key issue and concern is expressed about ensuring 'backup' is provided to cover days when there is little or no wind. However, in analysing this issue some studies make an explicit assumption that additional dedicated generating capacity must be built to 'firm up' wind and that this entails high additional costs to cover capital and operating expenses.

This is not a realistic assumption as, in practice, all fossil fuel and nuclear power stations provide backup to all others. As each has a statistical probability of experiencing an outage, more capacity is built than is required at peak demand levels to cover this eventuality. Wind is essentially no different although its availability is governed by the weather rather than the mechanical reliability alone. As wind is added to the system, it in itself, adds to system reliability and this is referred to as its 'capacity credit'. Importantly, as wind is added, it is not automatically the case that other power plants are retired. Analysis for the CCC show that use of flexibility introduced by storage, demand side response and interconnection, mean requirement for peaking plant, such as open cycle gas turbines for meeting shortfalls is low. The key point is that provision of full backup of renewable capacity is not necessary for secure supplies, where flexibility is encouraged.

Transmission

The cost of investment in transmission lines, cables and associated infrastructure is also a key theme, with Gibson and Civitas attributing very high transmission costs to wind. For simple situations, such as a wind farm on the end of line, analysis based on recent lifetime transmission costs suggest that costs are actually much lower. A difficulty in estimating cost of transmission on this basis is that transmission lines generally add to, or uprate, an existing interconnected system. The power flows are therefore more complex, lines are not loaded to maximum to ensure stability and security and there are often a series of related upgrades. Additionally, having insufficient transmission capacity necessitates the constraining of generation, which itself adds to balancing costs. In considering the cost of transmission expansion, it is important to note that other generation sources will also require transmission expenditure, not just wind. Mills and others emphasise that transmission expansion typically serves multiple purposes and that assigning the full costs of expansion to new (wind) generation capacity effectively ignores other benefits. The CCC suggest that transmission costs are likely to rise with renewables penetration. Wind generation in the north will tend to increase need for capacity but where it is closer to the south it may save costs accommodating non-renewable plant elsewhere.

Total 'system' costs

Substantial system costs exist, even in zero wind systems, precisely because the nature of electricity supply requires backup, balancing and transmission to allow individual, isolated, generators to contribute. The range of credible estimates from the literature for each component allows an estimate of total systems costs to be made for penetrations of up to 40% wind. These represent around 10% of the cost of onshore wind.

19.5   Life cycle carbon emissions

While wind power is low-carbon, it is not zero-carbon – carbon is emitted throughout the wind farm life cycle due to manufacture, construction, maintenance and decommissioning processes. The life cycle carbon emissions of wind power generation have been very widely studied, mostly by proponents of wind power who wish to demonstrate its low-carbon credentials, however, it is worth bearing in mind that life cycle carbon emissions estimates are not a complete picture and that there are also system effects of wind intermittency that may actually increase the carbon emissions of other parts of the network.

19.5.1 Calculation methodology

The life cycle carbon emissions of wind farms are conventionally calculated using partial process-based Life Cycle Assessment (LCA), defined by a number of national and international standards (BSI, 2011, ISO, 2006a, ISO, 2006b, ISO, 2013). This involves systematically analysing the greenhouse gas emissions of each process in each stage of the life cycle of the wind farm. There is considerable scope for variations to be introduced to the results by variations in assumptions, methodological choices and data uncertainty. Figure 19.4 illustrates the key areas where such variations might be introduced to estimates of the life cycle carbon emissions of onshore wind power.

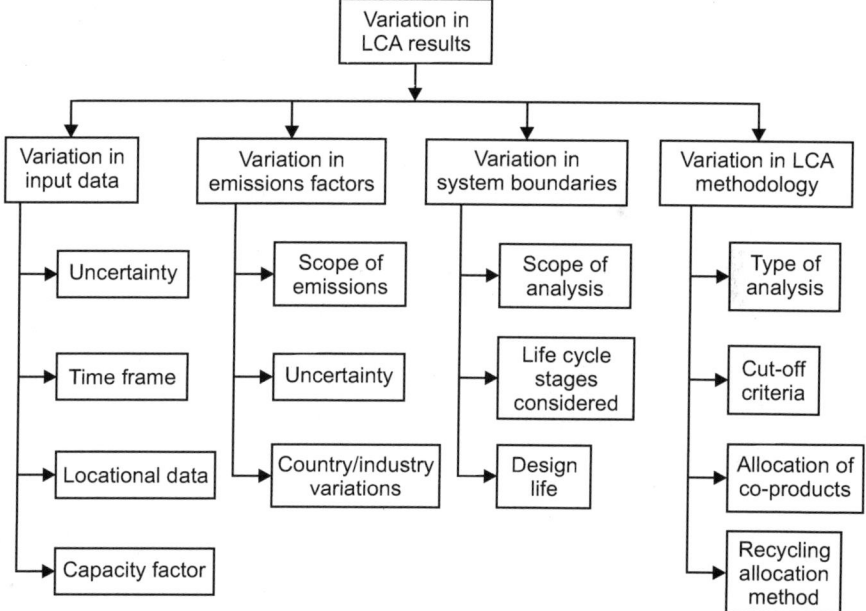

Figure 19.4: Causes of variation in LCA of wind farms.

These can be divided into four categories, which are described in greater detail in the main reports:

1. Variations in the input data stemming from variations in the wind farm scenarios, year, location and equipment, as well as uncertainty in this input data.

2. Uncertainties in the emissions factors extracted from life cycle datasets, variations in processes in different countries and differences in the scope of GHG emissions included.
3. Differences in physical and temporal system boundaries.
4. Variations introduced by differences in LCA methodology.

In a recent comprehensive review and harmonisation of carbon footprinting studies for wind power generation, the National Renewable Energy Laboratory (NREL) in the U.S. identified that the most significant variations in carbon emissions estimates were introduced by variations in capacity factor, but that there was generally a tight distribution of results, suggesting that 'new process-based LCAs of similar wind turbine technologies are unlikely to differ greatly'. In another review of the source of variation in carbon emissions estimates, Thomson identified that system lifetime and the methodology used for allocating recycling credits were also significant.

It is important to note, however, that capacity factor and system lifetime are a function of the specific design and location of each wind farm and are therefore likely to vary between farms. In arriving at their harmonised values found the mean assumed capacity factor of existing studies to be 30% onshore and 40% offshore, which are slightly higher than the respective values for the U.K. (27% and 39%). Good quality studies should include a sensitivity and uncertainty analysis to test the robustness of the results and their sensitivity to variations in methodology and key assumptions and estimates of life cycle emissions should be presented with uncertainty ranges.

Lifecycle impacts

The quality of published lifecycle carbon emissions estimates for on- and offshore wind vary widely, with only 41% of those identified by researchers at NREL meeting their basic quality screening criteria, therefore, only a selection of most robust and reliable estimates are considered here. Most of these studies apply process-based LCA, which is thought to suffer from truncation errors, but a small selection have been identified that are based on hybrid methods, which avoid these but may introduce double-counting errors – significantly there is no clear difference between the results from these two methods. The recent harmonisation published by NREL provides the most comprehensive review of published estimates of carbon emissions of wind power to date, finding mean estimates of 15 and 12 gCO_2eq/kWh for on- and offshore wind, respectively, following harmonisation of key assumptions.

The actual carbon emissions of onshore wind power generation in the U.K. might be higher than this, as the capacity factor is slightly lower than the NREL estimate and also some wind farms in the U.K. are constructed on peatlands, which was not considered in the NREL study. Peat plays a significant role in the carbon cycle and disturbance of the peat can lead to an increase in carbon emissions from the soil.

The manufacture and installation stages together account for over 90% of the total life cycle carbon emissions of an onshore wind farm not constructed on peatlands and 70% of an offshore farm, with the vast majority of these emissions arising during the extraction of materials and manufacture of components. Transport and installation typically contributes only about 6% of these emissions for an onshore wind farm if the carbon impacts of land-use change, such as construction on peatlands, are not included (which is common practice in existing studies). Offshore, the proportion will be higher as a result extensive use of vessels, although no study explicitly estimates the division between manufacture and installation impacts.

Of the remaining emissions, operation and maintenance activities contribute around 6% to the total life cycle impacts of an onshore wind farm and around 20% offshore (the latter being significantly higher due to the installation site being more challenging to access), while decommissioning accounts for a further 6%.

Comparisons with other generating technologies

Despite variations in estimated carbon footprint of wind power generation, it is significant to note they are all significantly lower than for fossil fuelled generation. Figure 19.5 compares the values presented here with those gathered by NREL for other types of generation, with the ranges showing the maximum range of published estimates. There is no overlap between wind generation and any type of fossil fuelled generation.

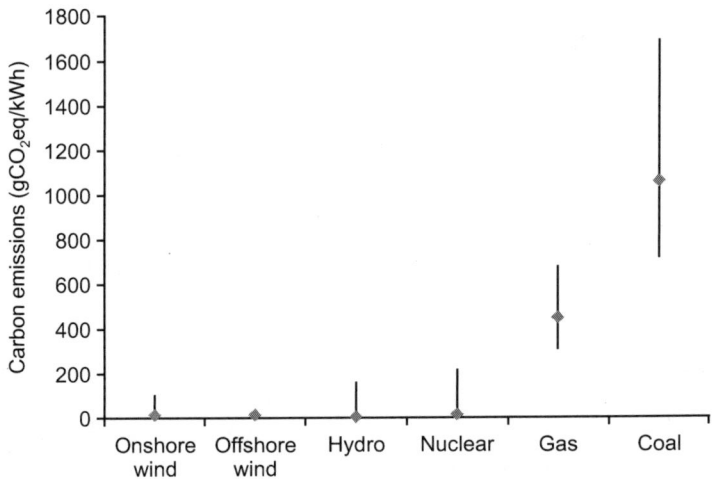

Figure 19.5: Comparison of carbon emissions of wind with other types of generation.

Furthermore, there is greater consensus on the carbon emissions of wind than there is for other forms of low carbon generation, such as hydro and nuclear power.

19.6 Carbon emissions displacement and payback of wind power

Estimates of the life cycle carbon emissions of wind farms are not, in themselves, particularly useful and are only really of interest for comparison with other forms of low-carbon generation. Further interpretation is required to calculate other values that may be more meaningful, such as the lifetime emissions reduction of a wind farm (the net reduction of greenhouse gas emissions taking into account both the life cycle carbon emissions and the lifetime emissions displacement) or the carbon payback period (the time for the emissions displacement to offset the life cycle carbon emissions). Estimates of carbon emissions displaced by wind power vary widely, as they are a measure of the displaced emissions resulting from wind power replacing other forms of generation. However, analysis of current research indicates that, while the carbon displacement of wind power generation can be approximated using such figures, this is likely to underestimate the positive impacts of wind power on carbon emissions.

19.6.1 Marginal emissions displacement

Accurate accounting of the emissions displaced by wind generation should reflect the fact that wind power only replaces certain types of generation (nuclear, for example, does not respond to fluctuations in wind). Generators operating on the margin, including conventional gas and coal power stations, operate at part load to provide reserve and respond to fluctuations in wind power and demand – the emissions displacement of wind power will be related to this marginal generating mix. There are also efficiency penalties associated with operating these power stations at reduced outputs, which causes the carbon emissions intensity of these generators to increase as a result of the presence of wind generation on the network.

Studies internationally confirm that the marginal emissions are significantly different from the system average emissions of the corresponding networks, with the actual values depending upon the types of generation available on the network and the relative prices of different fuels. The most relevant study specifically examined the recent marginal emissions displacement of wind power in Great Britain, taking efficiency penalties into account and found that, although efficiency penalties did reduce marginal emissions estimates, they remained higher than the corresponding system average emissions, for example, in 2012 the marginal displacement of wind power was 550g CO_2 eq/kWh, some 20% higher than reported, U.K., average emissions for that year.

Payback periods and lifetime emissions savings

The carbon payback period is the time for the carbon emissions displaced by wind power to equal the life cycle carbon footprint of the wind farm. In order to achieve a net reduction in GHG emissions, this should be significantly shorter than the intended lifetime of the wind farm. At current marginal displacement rates carbon payback is typically around 6 months to a year, although this can be several years for onshore farms built on peatlands where no effort has been made to mitigate the effects of wind farm construction.

Both the average and marginal emissions of electricity generation are likely to reduce over time, as the most polluting power stations are replaced with lower carbon alternatives. As such, the emissions displaced by a given wind farm will tend to decline over time, so that farms built further into the future will take longer to pay back. Pay back will be achieved as long as lifetime average emissions reduction factor exceeds the carbon footprint. Based on the most recent DECC forecasts for future grid emissions, Fig. 19.6 shows the lifetime average payback threshold for wind farms with a design life of 20 years, constructed between 2010 and 2050. It can be seen that the vast majority of carbon footprint estimates for onshore farms fall below this line and will

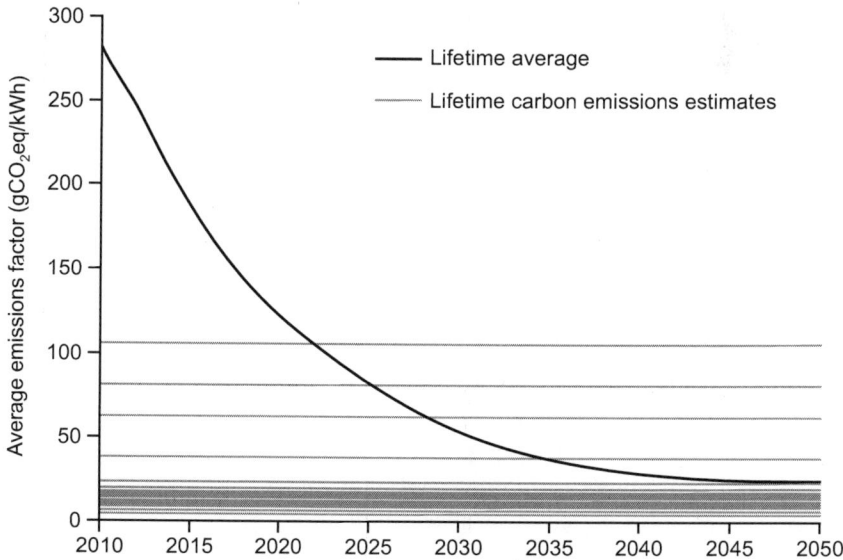

Figure 19.6: Carbon payback thresholds for wind farms constructed in the future, with estimates of emissions factor for onshore wind.

achieve carbon payback, however, the highest three, which correspond to onshore wind farms constructed on forested peat lands, will not achieve carbon payback if they are constructed after the mid-2020s. This is significant for wind farms currently being planned and highlights the importance of ensuring that undegraded peatlands are disturbed as little as possible. For offshore wind farms, all the current carbon footprint estimates fall below the payback threshold up until 2050.

To sum up, for life cycle carbon emissions the most critical aspects are the wind farm capacity factor, the system lifetime, the approach to recycling credits and uncertainty in emissions factors. Additionally, the emissions associated with construction on peatlands are important for onshore farms, while vessel use for installation and maintenance is significant offshore. While there is modest uncertainty associated with estimates of lifecycle emissions, values for both on- and offshore wind are substantially lower than unabated gas and coal generation and there are fewer inherent uncertainties than nuclear. Estimates of the lifecycle carbon emissions of offshore wind are generally lower than for onshore, due to better wind profiles and economies of scale.

Lifecycle carbon emissions also exclude 'system effects', as these are instead considered when examining carbon payback time or lifetime emissions savings, however the literature shows that, although efficiency penalties from operating thermal generation at part-load reduce the carbon savings from wind, the effect is modest. Furthermore, the rates currently used in wind farm analysis appear to systematically underestimate the emissions savings.

Wind generation is, therefore, effective at displacing fossil fuelled generation and reducing emissions, with carbon payback periods typically less than a year (although onshore construction on undegraded peatlands can extend this to several years). Long term, the expectation is that wind will remain effective at reducing emissions, even within an electricity system undergoing major decarbonisation. Furthermore, while wind farms on constructed on peatlands could soon reach the point that they would not be effective in reducing emissions, existing measures are available to minimise the impact on the peatlands so that this point can be pushed some distance in the future.

19.7 Future aspects of carbon footprint reduction in electricity generation technologies

Carbon footprints could be further reduced in all electricity generation technologies if the manufacturing phase and other phases of their life cycles were fuelled by low carbon energy sources. For example, if steel for wind turbines were made using electricity generated by wind, solar or nuclear plants. Using less raw materials would also lower life cycle CO_2 emissions, especially in emerging technologies such as marine and PV. New semi-conducting materials (organic cells and nano-rods), are being researched for PV as alternatives to energy and resource intensive silicon. Biomass has the potential to generate electricity with 'negative' CO_2 emissions. Burning 'carbon neutral' biomass and capturing the emissions using Carbon Capture and Storage (CCS) technologies would result in a net removal of CO_2 from the atmosphere. Studies show that a 'negative emission' of up to $-410gCO_2/kwh$ can be achieved. However, some researchers suggest that CCS, intended for large fossil-fuelled plants (>1000MW), would not be adopted for smaller capacity biomass plants, typically <50 MW.

Barriers to wind energy

20.1   Introduction

Wind energy today occupies a prominent position in the overall scheme of things in the power sector. From a technology that was regarded as fringe technology few years back, it has moved to center stage as an option offering near grid parity, which will become more attractive with scale. In many parts of the world, wind generates cheapest power at plant load factors as high as 40% or even 50%.

India, with nearly 22000 MW of wind power capacity ranks 5th in the world and very soon is likely to take over Spain in the 4th position. Capacity addition in 2018–19 is expected to be around 5000 MW. Indias target plan is to add 10000 MW, wind power installations every year from 2015–16 onwards.

Today, there are 20 wind turbine manufacturers in India with about 52 turbine models certified by National Institute of Wind Energy (NIWE) for grid connection. The total manufacturing capacity established in the country is about 10000 MW annually. Major wind turbine manufacturers are present in the country such as Vestas, GE, Gamesa, etc. having set up manufacturing plants in different parts of the country. The Indian market is dominated by the Indian Original Equipment Manufacturer (OEM), Suzlon which was the worlds 5th largest turbine manufacturer by the end of 2013. Suzlon acquired complete ownership of the German wind turbine manufacturer REpower in 2012 (renamed in 2012 as Senvion), becoming a major player in offshore wind power with turbines in the 5 MW+ series as well. Growth in wind power development has been energy shortage and even before the Rio conference when global warming and associated climate change got highlighted as a global environmental concern, windfarms were being set up in India.

20.2   Barriers and bottlenecks to scaling up wind energy

In the 12th plan period 2012–2017, renewable energy capacity of 32 GW is envisaged by Central Electricity Authority. Nearly 20 GW was to come from wind and 10 GW from solar. However, government plans to significantly scale up installations to almost 140 GW by 2020. 100 GW from solar and 40 GW from wind. This outlook also presents a new challenge on many fronts. According to industry estimates, if all the barriers to wind farm development are taken care of, nearly 40 GW (6–7GW/yr) of wind power installations could be possible by 2020. Such capacity addition and scale up, though not out of the realm of possibilities, however, requires a very detailed and critical examination of the barriers we face.

20.2.1   Key barriers

Some of the key barriers that have slowed down wind power development over the last 2–3 years are:

1. Highest merit order for wind in the generation scheduled (also known as the must run status) not being adhered to.
2. Merit order despatch for wind power is not observed uniformly throughout the country.
3. Inability of discoms to evacuate power in peak generation season and shut down of windfarms.
4. Fast tracking of green corridor implementation and strengthening of the transmission network not yet carried out.
5. Non availability of deemed generation benefits for grid loss.
6. Absence of centralised renewable energy management centre for better grid management and high RE infusion.
7. Payment default by discoms and inability to adhere to payment schedules for electricity generated from windfarms and significantly delayed payments up to six months.
8. Refusal to sign PPAs in time and in some cases in flagrant violation of approved norms.
9. Technical and administrative hindrances in allowing open access (encouraged in the electricity Act but discouraged by state governments).
10. Delays in project implementation due to right of way and land issues.
11. TARDY clearance process for forest land required for wind projects.
12. Complex and varied land related legislations hindering conversion of forest land.
13. Cross subsidy surcharge levels that have prevented sale to third party or even captive usage of windfarms - frequent changes in policies, practices and provisions.
14. Failure in implementation of RPO/REC and scheduling and forecasting regulations.
15. Lack of a common Renewable Energy Act solely for renewable energy projects which can exempt application of other laws of the land to wind power

These issues are discussed in detail in following broad areas.

20.2.2 Grid

A major barrier to wind power installations is the limitation of the grid, its limited power evacuation capacity and seemingly its inability to absorb very high penetration levels of wind power. Shutting down of large windfarms is a common practice adopted by Tamil Nadu utilities in the peak season when load demand in the region of windfarms is less than wind generation. Power evacuation system, which includes, sub-stations, transformers and transmission links are not adequate to carry the power generated from high wind areas to load centres. According to an estimate of Indian Wind Power Association (IWPA) a generation of around 2.0 billion kWh was lost because Tamil Nadu resorted to shut-down of windfarms in peak generation period. This is part of the problem. The other part is that frequent fluctuation in wind generation leads to frequent unscheduled inter-changes at the state level and from a grid discipline and grid security it becomes a major issue.

To achieve high penetrations of wind, one would not only need strengthening and expansion of grid but also operations and management would have to be completely revamped. Scheduling and forecasting of wind power generation from windfarms, a much debated topic, can resolve some of the problem but only to an extent. Additional storages (reservoirs and pumped hydro), Low Voltage Ride Through (LVRT) and High Voltage Ride Through (HVRT) capabilities, harmonic filters and the ability of the grid operator to adjust the generation from windfarms, keeping in view the grid conditions would also be required in order to address this problem in totality.

It is also interesting to note that in peak wind generation season, in almost every state, wind generation seems to have high correlation with energy consumption. If hydro storages or pumped hydro or other means of storage are deployed, variability in wind power output can be addressed. Industry experts are also of the opinion that variability in wind power output is not as big a problem or issue as it has been made out to be. At aggregate level with wind farms spread across vast regions extending almost over 500 km, the total variability on the power system is much less and much more predictable.

There is no doubt wind power creates challenges for the grid in many ways. A key problem in grid integration of wind energy is that it upsets the power balance that must exist between all the generating plants and the load demand. Any imbalance would affect the frequency of the system which could lead to loss of synchronism in certain cases. The accomplishment of power balance between the load and the generating plants is a major challenge. Another aspect that needs to be addressed is the continued operation of the windfarm in the event of a fault such as low voltage or high voltage or phase asymmetry. The system security comes under great risk if the large windfarm capacities drop out of the grid due to such faults. A capacity as large as 100 MW going out of the system can cause tripping of conventional generators and even complete breakdown. Therefore, modern wind turbines need to be grid compliant in the sense that the grid operator should be able to regulate power from wind turbines or windfarms, if needed. LVRT and HVRT kind of features in wind turbines will prevent large wind power capacities dropping out of the grid in the event of voltage disturbances.

The responsibility to resolve grid problems rests both with the utilities and the project developers and technology providers. In order to achieve the goals of scaled up capacity addition, these issues would have to be resolved by both in a phased manner. Renewable Energy incursion in the grid does throw up some formidable problems to the grid in the lower management on account of unpredictable wind power generation. However an effective and efficient way of Grid Management can definitely offset the minor challenges posed by wind power. Rather than thwarting wind power incursion what is required is a change of attitude on the part of utilities in the matter of wind power.

20.3   Policy uncertainty

The industry is still recovering from the recent experience of withdrawal of both AD and Generation Based Incentives (GBI). At the state level also there have been frequent changes in policies on open access, cross subsidy surcharge, banking and wheeling, group captive etc. These changes are often made with utter disregard towards projects that have been built and invested in based on certain assumptions of policy. Such changes can make businesses completely unviable. These changes in policies also have a detrimental impact on investors who are sitting on the fence to assess the overall investment environment. Often even announcements such as those of competitive bidding by a state can have an impact on investments. There are policies that are declared and announced by the State governments but there are also policies that are not announced but practiced in silence. Refusal to allow open access to generation projects within the state or a heavy cross subsidy burden on third-party sales are such state policies which are not publicised but are implemented. Such curbs have slowed down project development at state level significantly.

There is a need for uniform policies across the country as much as possible and these policies should be applicable over a reasonable period of time (5 years) to achieve fruitful results. Changes in tariffs and any other charges should be transparent and there should be a long-term visibility.

Wind projects should be considered as critical assets or as prohibited area to minimise thefts and related damage as the installations are located in inhospitable/sparsely populated far flung areas. Wind power projects need to be exempted from obtaining 'Consent to Establish' from individual departments as the allotments are made by the single window Nodal Agency of the states.

There is a specific need for separate legislative enactment in the form of Renewable Electricity Act to be in place in the quickest possible time.

20.4   RPO compliance

Renewable Purchase Obligation (RPO) compliance has not been strictly enforced or properly monitored, directly affecting the REC mechanism. Although many projects have been registered under the REC registry and an increasing no. of RECs has been issued, the prices have been low with a majority of RECs being sold at the floor price or not sold at all. RPO compliance made binding will immediately result in much relief to a large number of IPPs and will pave the way for more investments.

1. Enactment of Renewable Energy Act will mitigate to some extent the difficulties of RPO implementation.
2. States should fulfil RPO in line with NAPCC targets.
3. REC offtake should have a good incentive formula for fillip in the market.
4. Procurement of REC by a central agency at CERC determined tariff.

20.4.1   Wind park/central procurement

One of the ways of enhancing capacity addition in windfarms and also investments in sector is to identify large wind zones for wind parks or wind-solar hybrid parks that are initially developed by the state government or in PPP mode. Electricity generated can be procured by a central government agency and then made available to state governments. Such projects can be connected to Power Grid Corporation of India Limited (PGCIL). Wind parks or wind-solar hybrid parks with central procurement of electricity and initial development by the state government could be one of the ways to minimise risks in wind power projects and to enhance investments.

Repowering many of Indias wind farms were set up during the early 90s. Thus we see that for some of the best wind power sites in India, older turbine models of smaller rotor diameters and lower hub heights with lattice structure towers were used. Wind turbine technology and understanding of wind farm design, micrositing and analysis have evolved and progressed since then. New WTG models are able to reach higher efficiency levels due to better rotor and airfoil designs. IEC Class III turbines with larger rotor diameters and better control systems allow for increased energy capture and turbine reliability. Taller towers enable capture of stronger, less turbulent wind at greater heights. Keeping all this is mind, an active repowering strategy and policy is very much required in the country. Repowering of old wind farms would lead to significantly increased annual energy yield from the same patch of land where old wind turbines are functioning. It is assessed that repowering potential is of the order of 5000 MW and with the passage of time more and more wind farms could possibly qualify for repowering.

20.4.2 Land availability

Many areas suitable for wind farms are notified as forest areas. However, actually all the notified forest areas are actually not forests. Many areas are shrub lands with some cattle grazing that make these suitable for wind power development. There is a need to analyse the land use land cover data of India to identify the regions most suitable for wind farm development. The process of making these patches of land dedicated for windfarms should be made simpler. Also approvals and clearances such as land title conversion should be examined in detail to find ways of streamlining acquisition and usage of lands for the development of wind farms.

1. Easing of forest area land acquisition formalities and expediting the same.
2. Exemption from wildlife zone applicability, migratory bird route limitations, air flight path restrictions. Not to be applicable for wind power.
3. No ceiling limitation for land acquisition under RE projects.

20.4.3 Local development/employment

Wind power development should also lead to local development and employment generation. Project developers should keep in mind that as long as project development is aligned with the well being and progress of local population in the vicinity of the wind farms, problems such as Right of Way issues and law and order problems are unlikely to happen. Innovative ways of involving villagers in windfarm ownership and making them a stake-holder in the project should be devised. Perhaps CERC should determine a different tariff for projects that have community participation. Community windfarms are a worldwide movement and India should find its own unique solutions to community participation. Government, regulators and project developers need to examine these aspects.

20.5 Financial issues

20.5.1 Debt and equity barriers

1. Limited long term debt market:
   - Under developed bond market.
   - Asset liability mismatch.
   - Limited takeout financing options.

2. High financial cost:
   - High inflation rates.
   - Competition among multiple sectors.
   - Uncertainty about countrys future borrowing needs.
3. Volatile market:
   - Fluctuation in inflation rates.
   - Interest rate fluctuation.
   - Uncertain policy environment.
4. Lack of fixed interest rate loans:
   - Unavailability of long term hedging instruments.
   - Uncertainty in financial markets.
5. Availability of funds:
   - Variability of wind power and skewed regulations.
   - Capping on ECB.
   - Under-developed bond market.
6. Interest rate ceilings on ECB:
   - Higher risk perception with wind makes it unattractive.
   - High hedging cost restrictions to provide funds.
7. Lack of forwards market in foreign exchange.
   - Nascent forward market.
   - High cost of hedging in India.
8. Poor financial health of distribution utility:
   - Most of the WPPs executed PPA with discoms.
   - Poor condition of discoms and delayed payments increase uncertainty.
9. Lack of refinancing options:
   - Cashflow reduction in initial years, reduce chances for refinancing.
   - Increasing funds availability by making REC a separate priority sector and creation of debt funds.
   - Reducing funds cost by tax free RE Bonds and increasing debt period to project life cycle.
   - Reduction in MAT rate to 10%.
   - Covering power sector under GST and at 0 tax.
   - 80IA benefit till 2020.
   - Incentives for export of WTGs.
   - Reducing overheads such as taxes, duties for open access transactions.

20.6 Manufacturing capacity

The total manufacturing capacity in India is reported of the order of 10000 MW/year. Since the maximum capacity installed in recent years was only around 3200 MW, most of the manufacturing capacity remains partially or un-utilised. It is felt that if wind turbines of the order of 7–10000 MW are required to be manufactured in the country, it may not be difficult to do so. However, it is important also that as much as possible, cutting edge technologies are deployed. Since many of the wind turbine suppliers or OEMs in India are manufacturing sub-megawatt class wind turbines that appear to be obsolete in other parts of the world, we may have some bottlenecks.

20.6.1 Bankable project

Developing bankable projects takes about three years. Bankable sites require that wind speeds be measured at a site for at least one year. To arrive at one feasible site, a developer may have to measure wind speeds at three sites. This aspect is going to create bottlenecks as there must be a limit to the sites that have measured wind speed data in accordance with international standards and are also feasible. This will be yet another bottleneck in scaling up wind power development. The practice of some SERCs to announce the tariff on yearly basis goes against the grain of the three year bankability for Wind Power Projects. By the time the developer decides to install the project after getting all the clearances the tariff goes for the toss which tips the scales of Economy and viability of project. Tariffs should be long term and transparent. Frequent changes in Tariff structure numbs the pace of investment as it amounts to shifting of goal posts.

20.6.2 Supply chain issues

There are a few companies that manufacture gearboxes, generators, rotor blades and control panels for wind turbines. Some OEMs do not need gear boxes and they make their own generators. Similarly some OEMs manufacture their own rotor blades. Transformers, instruments, bearings and shafts and the tower parts are also supplied by a few vendors. In many cases the same vendor supplies components for different OEMs. It is possible that due to the increased demand, the prices of these components increase and so do the delivery periods. The same holds for some of the raw materials like steel for towers, shafts, nacelle frame, fibreglass, magnets for permanent magnet generators, etc.

20.6.3 Project implementation

Equipment needed for handling of heavy wind turbine components and parts and for transportation and assembly at site will also pose bottlenecks to project development. Cranes to set up wind turbines at 100 m or 120 m towers are also going to be scarce. As we look towards development of 40 GW capacity by 2020, we find that there are many areas where policy intervention is needed and many areas where bottlenecking is possible. These issues are summarised in the Table 20.1.

Table 20.1: Constraints or bottlenecks in wind sector in scaling up the installations.

Grid	*Sub-stations*
Policy uncertainty	Transmission links
	Transformation capacity
	Scheduling and forecasting
	Need to have long-term visibility
	Center
	State
	Matters such as cross-subsidy surcharge etc.
RPO	Should be legally binding
Repowering	Model, framework and policy needed
Hybrid wind – solar parks	Model, framework and policy needed
Land availability	The process of land acquisition be made simpler
	Why is title conversion needed
	Forest lands that are not forests
Local development and employment	Developers to be sensitive towards this need
	Perhaps some policies by the government
Manufacturing	Bottlenecking possible in case of cutting edge technology
Bankable projects	Bottlenecking possible
Supply chain	Bottlenecking and cost increase possible
Project implementation	Availability of heavy duty handling equipment such as cranes could be a constraint
Tariff	Should be fixed for long terms and frequent change of Goal post should be avoided

20.7 Barriers and incentives to wind energy development

20.7.1 Institutional barriers

1. Unfair competition: Compared with wind energy, nuclear and fossil fuel technologies enjoy a considerable advantage in government subsidies for research and development.

2. Internalisation of generating costs: Wind energy will be unable to compete on a level playing field with conventional generation until new policies are adopted to internalise the public costs of these fossil and nuclear fuel sources.

3. Lack of knowledge: Local electricity companies may be unfamiliar with wind energy. Most utilities have not studied how renewable resources could fit into their systems. For example, few have investigated how the output of wind technologies matches their system peak load.

20.7.2 Regulatory barriers

1. Interconnection standards: There is a lack of uniform interconnection standards for wind technologies. The responsibilities of utilities and generators with respect to interconnection have not been clearly defined and as a result, the cost (or even feasibility) of interconnection to the grid often becomes a significant barrier to smaller projects.

2. Licensing requirements: Stringent licensing requirements could also pose a barrier if for example, a small cooperative selling power only to its members were required to be licensed as an energy service provider. Similarly, the licensing of generation facilities could be particularly onerous for small developers.

3. Unfair competitive disadvantage: Wind often face an unfair competitive disadvantage because public policies do not generally fully account for the environmental and social costs of conventional electricity supply technologies.

20.7.3 Investment barriers

1. Lack of a model: Since there are few wind project in the region, project developers are understandably unfamiliar with wind power and even less comfortable with the risks involved.

2. Difficulty raising local equity: Local developers have limited investment capacity and there are many projects competing for investment capital. As an unfamiliar and potentially risky investment with uncertain returns, wind power does not immediately emerge as a clear winner.

3. Obtaining capital resources: Proper risk allocation and mitigation are fundamental concerns for wind developers seeking to secure project funding. In a typical wind project, there are common barriers to accessing capital resources, such as the availability of 'financeable' offtaker agreements, construction cost over-runs, longterm reliability of turbines and the risk of wind-bearing ability.

20.7.4 Technical barriers

1. Wind resource assessment: The wind maps developed by some countries are not enough for project developers. High resolution wind maps are needed.

2. Intermittency: Because the wind blows intermittently, wind power can only supply a portion of a consumers, communitys, or regions electricity needs. In grid-connected applications, owners of wind capacity must negotiate supply contracts that account for winds intermittent nature.

3. Wind in rural areas: Rural small grids would require upgrading to three-phase before interconnecting wind turbines larger than 20–25 kW. Without standard agreements concerning each partys responsibilities in paying for necessary upgrades, this weakness in the grid is a significant barrier to wind development in rural areas.

20.7.5   Commercialisation barriers

1. Infrastructure: Developing new wind sites requires large initial investments to build infrastructure. These investments increase the cost of providing wind electricity, especially during early years.

2. Prospecting: Developers must find publicly acceptable sites with good resources and with access to transmission lines. Potential wind sites can require several years of monitoring to determine whether they are suitable.

3. Permitting: Planning permission issues for conventional energy technologies are generally well understood and the process and standards for review are well defined. In contrast, wind energy often involve new types of issues.

4. Marketing: Individuals have no choices about the sources of their electricity. But electricity deregulation could open the market so that customers have a variety of choices.

20.7.6   Market barriers

1. Small size: Renewables projects and companies are generally small. Thus they have fewer resources than large generation companies or integrated utilities. They will have less clout negotiating favourable terms with larger market players. And they are less able to participate in regulatory or legislative proceedings, or in industry forums defining new electricity market rules.

2. High transaction costs: Small projects have high transaction costs at many stages of the development cycle. For example, it costs more for financial institutions to evaluate the credit-worthiness of many small projects than of one large project.

3. High financing costs: Wind developers and customers may have difficulty obtaining financing at rates as low as may be available for conventional energy facilities.

4. Lack of information: Financial institutions are generally unfamiliar with the new technologies and likely to perceive them as risky, so that they may lend money at higher rates.

5. Transmission costs: Wind projects may also be charged higher transmission costs than conventional technologies or may be subject to other discriminatory grid policies.

20.7.7   Instruments to overcome barriers

1. Goals and objectives have to be established at a regional level based on the experience that individual countries have developed up to know.
2. Having established its goals and objectives, decision makers can design policy tools that seek to fulfill them.
3. Regulatory reforms should be considered to ensure that wind energy project could feed into the power grids. Further, policy makers may elect to adopt targeted measures that address specific barriers to wind energy and promote investments in such systems.

20.7.8   Market incentives

1. Wind energy targets: The interconnected system could guarante minimum percentage of wind energy to be part of the overall energy supply portfolio. The recently approved PROINFA Programme in Brazil follows this approach by requiring that the government owned utility - electrobras - purchase a minimum amount of renewables-based electricity by a certain date.
2. Specific subsidies for wind energy: This system is basically a tax collected from all power services, which goes into a fund to support wind energy developments. The U.K. has had a variation of this with its competitively bid 'Non-Fossil Fuel Obligation' programme. The government imposes a levy on all retail electricity sales to help finance renewable energy projects.

20.7.9   Tax incentives

1. Tax-free generation: Tax-free generation is found primarily in Sweden and Denmark, where each member of a cooperative, commune, or partnership is not taxed on his or her share of the income from the turbines production as long as the income (or the amount of electricity produced) does not exceed that members annual expenditure on (or consumption of) electricity.
2. Favourable depreciation: Businesses are allowed to depreciate the value of a wind turbine by up to 30% each year using the declining balance method and to use the depreciation expense to offset other forms of business income Swedish and Danish farmers have installed one or more wind turbines on their property as part of their farming business – to defer taxation on their other farming profits.

20.7.10 Standard interconnection agreements

1. Distribution utilities should be required to interconnect wind projects to the grid according to a predetermined set of rules defining technical requirements and division of financial responsibility.
2. Pre-defining interconnection requirements and responsibilities (both technical and financial) enables a wind project to accurately estimate the cost of interconnection in advance.
3. These factors reduce the project owners risk.

20.7.11 Wind turbine manufacturing base

1. World-class wind turbine manufacturing industry has played an important role in developing wind energy in many countries. The very existence of a domestic wind turbine manufacturing industry has no doubt influenced politicians in those countries to favour policies that are friendly towards domestic ownership.
2. Brazil has followed this path with an incipient but strong wind turbine manufacturing industry.

20.7.12 Tariff schemes

1. Voluntary approaches like green tariffs and green shareholder programmes are based on a high consumers' willingness to pay for 'green electricity'.
2. Feed-in tariffs: It guarantees to all wind energy producers up to certain level of the current domestic sale price of electricity for every kilowatt hour they generate. This is both administratively simple and effective in practice mechanism for promoting wind energy. It has created a stable, profitable and essentially unlimited market for wind power and one that can be accessed with very low transactions costs.

Three key levels of strategic planning:

1. Electricity market.
2. Electricity network.
3. Spatial planning.

They need to be integrated into a plan led approach to wind energy deployment.

Economics of wind energy

21.1   Introduction

Wind energy capital costs have declined steadily. The corresponding electricity costs vary due to wind speed variations, locations and different institutional frameworks in different countries.

At low fossil fuel prices, wind energy has not generally been cost competitive with the thermal sources of electricity generation. The pattern of development of wind energy is largely dependent on the subsidies and support mechanisms provided by national governments.

21.2   Economic considerations

The big driver behind the growth in wind energy investment is the falling cost of wind-produced electricity. Higher fossil fuels such as oil, coal and natural gas prices are helping to make wind power more competitive. Even where wind power is still not able to compete head-on with cheaper power sources in some locations, it is getting close.

21.2.1   Transmission and grid issues

Transmission costs are a major issue in wind energy development. Some of the best locations for generating wind energy are far distant from the consuming industrial and population centers. Some areas have a better, more reliable source of wind power than others. Although half the U.S. installed wind power capacity is based in Texas and California, the greatest potential for wind generation can be found in areas where there is little demand for electrical power. For instance, there exists a significant amount of wind potential in North Dakota, but there are just not a lot of people or industries in North Dakota to consume the electrical power. The highest wind speeds exist in the remote and inaccessible Aleutian Islands in Alaska and necessitate an energy storage and conveyance medium such as hydrogen from water as a transportation fuel in fuel cells. A massive upgrade of the transmission lines nationwide through the national electrical power grid using high voltage DC instead of high voltage AC is needed to tap those distant sources.

Where water supplies are abundant, along sea shores or internal lakes or rivers, the electricity produced could be used for extracting hydrogen from water through the electrolysis process. Hydrogen then can become the storage medium and energy carrier of wind energy. It would be conveyed or transmitted to the energy consumption sites possibly through the existing natural gas pipeline system which covers the U.S.

Another alternative is to convert hydrogen with coal into methane gas, (CH_4) that could be distributed through the existing natural gas distribution grid without significant modifications. Methane itself can be converted into methanol or methyl alcohol, (CH_3OH) as a liquid transportation fuel. In the long term, to reduce the electrical transmission losses, one can envision superconducting electrical transmission lines cooled with cryogenic hydrogen carrying simultaneously electricity and hydrogen from the wind energy production sites to the consumption sites. Such a visionary futuristic power transmission system could also provide the electrical power for a modern mass transit system using Magnetically-levitated (Maglev) high speed trains transporting goods and people supplementing the current highway system in the U.S.

21.2.2   Wind electricity cost and price

An important characteristic of wind energy production is that there is no such thing as a point price for wind energy. A linear relationship does not mathematically exist. This is so since the annual electricity production will vary largely depending on the amount of wind available at a given wind turbine site. Thus a single price for wind energy does not exist, but rather a price interval or range, depending on wind speeds.

21.3   Wind energy cost analysis

In general, generation technologies, the cost of electricity is primarily affected by three main components:

1. Capital and investment cost.
2. Operation and maintenance (O&M) cost.
3. Fuel cost.

However, studies of the cost of wind energy and other renewable energy sources could become flawed because of a lack of understanding of both the technology and the economics involved. Misleading comparisons of costs of different energy technologies are common. It is misleading to think that the amount of funds needed to pay for the purchase a wind turbine is a cost or expenditure. Even the realised profit cannot be considered as a cost.

Specifically, the cost of electricity in wind power generation includes the following components:

1. Economic depreciation of the capital equipment.
2. Interest paid on the borrowed capital.

3. The operation and maintenance costs.
4. Taxes paid to local and federal authorities.
5. Government incentives and tax credits.
6. Royalties paid to land owners.
7. Payment for electricity used on a standby mode.
8. Energy storage components, if used.
9. The cost of wind as fuel is zero.

21.3.1 Levelised cost of electricity (LCOE)

The Levelised Cost of Electricity (LCOE) in electrical energy production can be defined as the present value of the price of the produced electrical energy in cents/kW.hr, considering the economical life of the plant and the costs incurred in the construction, operation and maintenance and the fuel costs.

Discount rate

The discount rate, is chosen depending on the cost and the source of the available capital, considering a balance between equity and debt financing and an estimate of the financial risks entailed in the project. It is advisable to consider the effect of inflation and consequently using the real interest rate instead.

Net present value

The net present value of a project is the value of all payments, discounted back to the beginning of the investment.

Real rate of return

The real rate of return is the real rate of interest 'r' which makes the net present value of a project exactly zero. The real rate of return is a measure of the real interest rate earned on a given investment. The computation of the real rate of return requires an iterative procedure to find the roots of the expression for the present value.

Electricity cost per unit energy

The electricity cost per kW.hr is calculated by first estimating the sum of the total investment and the discounted value of operation and maintenance costs in all years. The income from electricity sales is subtracted from all non-zero amounts of payments at each year of the project period.

Depreciation cost

Depreciation is a term used in accounting, economics and finance to spread the cost of an asset over the span of several years. In simple terms, it can be said that depreciation is the reduction in the value of an asset or good due to usage, passage of time, wear and tear, technological outdating or obsolescence, depletion, inadequacy, rot, rust, decay or other such factors.

We cannot calculate the economic depreciation of an investment unless we know the income from the investment. Depreciation is defined as the decline in the capital value of the investment using the internal rate of return as the discounting factor. If the income from the investment is not known, the rate of return is not determined, thus one cannot calculate economic depreciation.

The tax depreciation or accounting depreciation is sometimes confused with economic depreciation. However, tax or accounting depreciation is a set of mechanical rules which must not be used when the true cost of energy per kW.hr is sought.

Straight line depreciation: Straight line depreciation is the simplest and most often used method in which we can estimate the real value of the asset at the end of the period during which it will be used to generate revenues, or its economic life. It will expense a portion of the original cost in equal increments over the period. The real value is an estimate of the value of the asset or good at the time it will be sold or disposed of. It may be zero or even negative.

21.4   Price and cost concepts

The words: 'cost' and 'price' are sometimes mistakenly used as synonyms. The price of a product is determined by supply and demand for the product. Some people assume that the price of a product is somehow a result of adding a normal or reasonable profit to a cost, which is not necessarily the case, unless it is applied to a government controlled monopoly. Thus in general:

$$\text{Price} = \text{Cost} + \text{Profit} + \text{Taxes} + \text{Installation} + \text{Fuel}$$
$$+ \text{Operation and maintenance}$$
$$- \text{Government incentives and tax credits}$$
$$+ f(\text{scarcity})$$

A factor that is a function of the scarcity of wind resources at a given location, as well as the taxes paid, must be accounted for.

21.4.1   Wind turbines prices

Wind turbine prices may vary due to transportation costs, different tower heights, different rotor diameters, different generators sizes and the grid connection costs. It must be realised that some of the manufacturers deliveries are complete turnkey projects including planning, turbine nacelles, rotor blades, towers, foundations, transformers, switchgear and other installation costs including road building and power lines. The manufacturer sales figures also include service and sales of spare parts.

The manufacturers sales include licensing income, but the corresponding rated power in MWs are not registered in the company accounts. Sales may vary significantly between markets for high wind turbines and low wind turbines. The prices of different types of turbines are quite different. The patterns of sales, types of turbines and types of contracts vary significantly from year to year and depend on the different locations and markets.

Productivity and costs

A unique aspect of wind energy production is that its productivity and costs depend on the price of electricity and not *vice versa* as in other energy systems.

The annual production per m^2 of rotor swept area in a location like Denmark tends to be significantly higher than in another location such as Germany. This has no relationship to the different wind resources. It is instead related to the different prices for electricity at the different locations. In Denmark it is not profitable to locate wind turbines in low wind areas, whereas it is profitable to use low wind areas in Germany due to the higher electricity prices.

Germany has a very high electricity price for renewable sources of electricity in terms of the tariff per kW.hr of energy delivered to the grid.

In Germany it is profitable to equip wind turbines with very tall towers. The high electricity price also makes it profitable to locate wind turbines in low wind areas. In that case, the most economic turbines will have larger rotor diameters relative to the generator size than in other areas of the world.

Wind turbines sold in the German market appear more expensive than they do in other markets, if one considers the price per kW of installed or rated power. However these are machines optimised for the German low wind sites. The price per square meter of rotor swept area located at a given hub height is what matters, not the price per kW of installed power.

Installation costs

Another unique feature of wind energy is that a high cost of generating electricity is not necessarily a result of high installation cost. One incurs a high installation cost whenever a good wind resource is available and hence cheap generating costs are available in a remote area.

High installation costs can be afforded, typically when a good wind resource exists since the power produced by a wind turbine is proportional to the cube of the wind speed. The installation costs include the costs for extension of the electrical grid and the grid reinforcement. The costs of electrical cabling can be significant, affecting whether a wind farm is located next to an existing medium voltage power line or far from a power line.

Average installation costs cannot be used, since the electricity price per kW.hr delivered to the grid depends upon the distance to the grid. Installation costs may vary with the location, road construction and grid connection amounting to about 30% of the turbine cost.

Operation and maintenance

The operation and maintenance cost can be estimated as either a fixed amount per year or a percentage of the cost of the turbine. This could also include a service contract with the wind turbine manufacturer.

Location effect

One cannot use the statistics from a specific area to estimate the costs in another area. The cost of wind energy in Germany is high, because prices for electricity are high and the cost of wind energy in the U.K. is low, because the price of electricity is low. Few wind turbines will be installed if the price of electricity is low, because high wind sites are scarce and sites which are profitable may not be found.

Price per square meter of rotor swept area

The price per kW of rated power is not a good guide to investment in wind energy project. What really matters is the price per square meter of rotor swept area. This is analogous to farming where the price per acre of farmland is the relevant capital expenditure in addition to the price of the machinery and farm structures.

The price of a wind turbine per kW installed power is usually difficult to get hold of and a very poor guide to cost developments. It is difficult to give a single figure for the price per kW of installed power, simply because the price of a turbine varies much more with its rotor diameter than with the rated power of its generator. The reason is that the annual production depends much more on the rotor diameter than the generator size. The comparisons of average price per kW of installed power for different energy technologies are usually misleading, if wind power is considered.

Intermittence factor

The Intermittence Factor (IF) or Capacity Factor for an energy generating technology is equal to the ratio of the annual energy production to the theoretical maximum energy production, if the generator were running at its rated electrical power all year.

$$\text{Intermittence factor IF} = \frac{\text{Annual energy production}}{\text{Rated maximum energy production}}$$

Depending on the wind statistics for a particular site, the ideal capacity factor for a wind turbine is in the range of 25–40%, because that capacity factor minimises the cost per kW.hr of energy produced. An apparent paradox is that it is not desirable to increase the capacity factor for a wind turbine, as it would be for technologies where the fuel is not free. Capacity factors will be very different for different turbines, but likewise the prices or costs of these turbines will be very different. Overall, what counts is the cost per kW.hr of energy produced, not the capacity or intermittence factor.

Land rents, royalties and project profitability

In wind energy production the land rents or royalties should depend on the profitability of a project and not *vice versa.*

The compensation, land rents or royalties paid to land owners where the turbines are placed is sometimes treated as a cost of wind energy. In fact, it is only a minor share of the compensation which is a cost of the loss of crop on the area that can no longer be farmed, a possible nuisance compensation since the farmer has to make extra turns when plowing the fields underneath the wind turbines and he must be compensated for compaction and the damage to tiling from the heavy equipment access to the turbine site. If the compensation exceeds what is paid to install a power line pylon, the excess is in fact an income transfer. This is a different matter economically: it is not a cost to society, but a transfer of income or profits from the wind turbine operator to the land owner. Such a profit transfer is called a land rent by economists. A rent payment does not transfer real resources from one use to another.

There is no standard compensation for placing a wind turbine on agricultural land. It depends on the quality of the site, the availability of the wind and the grid access nearby. A land owner can bargain for a high compensation in a good location, since the turbine operator can afford to pay it due to the profitability of the site. If the site has low speed wind and high installation costs, the compensation will be estimated closer to the nuisance value of the turbine.

Project lifetimes

The figure used for the design lifetime of a typical wind turbine is 20 years. With the low turbulence of offshore wind conditions leading to lower vibrations and fatigue stresses, it is likely that the turbines can last longer, from 25–30 years, provided that corrosion from salty conditions can be controlled.

Offshore foundations for oil installations are designed to last 50 years and it may be possible to consider two generations of turbines to be built on the same foundations, with an overhaul repair at the midlife point after 25 years.

Payments

The payments, including the initial payment, are used to calculate the net present value and the real rate of return over a 20 years project lifetime since this is the main economic aspect of the analysis. The tax payments and credits and the depreciation credits are not considered for simplification but could be added for a more detailed analysis later. We consider that the capital is in the form of available invested funds. If the capital cost is all borrowed funds, then the interest payment on the loan or the bonds must be accounted for.

Operation and maintenance: 1.5% of turbine price

Total expenditure = Total turbine cost + Operation and maintenance cost over expected lifetime.

It will always be argued that the fuel for a small wind turbine is free, there is not much maintenance required, nor personnel to run it. Its operational costs are minimal. A small wind turbine energy production does not need power lines to be delivered to the customer if they are already there in his backyard. No costs for the use of electricity networks and no grid losses are incurred. These could reach the 10% that conventional electricity looses during transport and distribution. No administrative or overhead costs are incurred. On average, the sale price of traditionally generated kWhrs is about 10 times the depreciation costs. If these factors are taken into account, the costs comparison between a small wind turbine kWhrs and conventionally generated electricity start converging toward each other.

It must be admitted that the electricity from wind turbines is currently more expensive than traditional generation such as from coal or natural gas. However the future scarcity would bring cost increases as rising fuel costs, wage hikes and environmental requirements will affect the costs of conventionally generated electricity, but not wind energy's.

If one shares the view that the stock of fossil energy is finite and depletable and that the first signs of shortage are already appearing on the horizon, then one is compelled to recognise the threat of international vicious competition for the control of the remaining supplies of fossil fuels. This makes a compelling case for wind energy.

Offshore wind farms economics

The most important consideration why offshore wind energy is becoming more economical is that the cost of building the foundations has significantly decreased. The estimated total investment, including the grid connections, required to install 1 MW of wind power offshore in Denmark is around 12 million Danish Korona (DKK), equivalent to 4 million German Marks (DEM), or $1.7 million/MW. Since there is substantially more wind at sea than on land, the average cost of electricity at 5% real discount rate and a 20 years project lifetime is about 5 cents/kWhr including an operation and maintenance cost of 1 cent/kWhr.

Project lifetime

One would think that turbines at sea would suffer corrosion from sea water leading to a shorter lifetime. However winds at sea have a lower turbulence than winds onshore leading to lower vibrations and resulting in a longer lifetime for turbines at sea. Assuming an extended project lifetime of 25 years instead of 20, this leads to costs that are 9% lower.

Danish power companies have been optimising their wind projects with a project lifetime of 50 years. They require a 50 year design lifetime for the foundations, towers, nacelle shells and main shafts in the turbines.

If the turbines have a lifetime of 50 years, they will require an overhaul or refurbishment after 25 years. That should cost some extra 25% over the initial investment. This leads to a cost a cost of electricity of 0.283 dkk/kWhr, which is similar to the one for average onshore locations in Denmark.

Fishing industry analogy

Wind power production has unique characteristics that are different than other energy production systems. It bears up a close analogy to the fishing industry, in that to catch the fish that is available for free from the oceans, capital expenditures are needed to purchase a boat and the fishing nets, operational costs are needed to maintain the boat and its equipment in a running condition, human labour is needed to operate the boat and then most important, markets are needed to sell the fish catch.

Similarly, to harvest the wind that is available for free from the air, capital expenditures are needed to purchase the wind turbine and the associated power conditioning and transmission equipment.

Operational costs are needed to maintain the turbine in a standby mode ready to catch the wind when it starts blowing, extracting power from the grid at a low level of 2–5 kW to operate its control and ventilation equipment. Human labour is needed for control and maintain the turbine components. Connection to the grid to sell the excess power produced is then needed when the wind blows favourably.

A unique characteristic of wind power production is that the cost incurred and the associated price that can be charged for the produced electricity are not constant and decrease as the total amount of wind electricity is increased.

Countries that have a history in the fishing industry, like Denmark have grasped this reality and have encouraged the individual and cooperatives ownership of wind turbines, much like the ownership fishing boats by individuals and small businesses and provided them with a market for the product electricity by passing laws requiring the utility grid to purchase from them the product electricity. The system is so successful that Denmark has been exporting its surplus wind electricity to the European Union.

In contrast, under the antiquated protected utilities monopoly system in the U.S., the concept of net-metering allows individuals to only get credit of their excess wind electricity production up to the amount of electricity that they purchase from the grid and a lower price than the electricity purchased from the grid. Any credits for excess power production are forfeited by the utility at the end of the year.

This discourages the production and export of excess wind electricity and other renewables, into the grid system and amounts to an amateurish catch and release of the excess captured power, denying potential producers the benefit of being able to market their excess power production. Such a hurdle must be overcome for a sustainable and meaningful implementation of wind power production in the U.S.

21.4.2   Factors affecting cost of wind energy

Thus, the factors affecting the cost of wind energy are still rapidly changing and wind energy's costs will continue to decline as the industry grows and matures. A number of factors determine the economics of utility-scale wind energy and its competitiveness in the energy marketplace.

Cost of wind energy varies widely depending upon the wind speed at a given project site

The energy that can be tapped from the wind is proportional to the cube of the wind speed, so a slight increase in wind speed results in a large increase in electricity generation.

Improvements in turbine design bring down costs

The taller the turbine tower and the larger the area swept by the blades, the more powerful and productive the turbine. The swept area of a turbine rotor (a circle) is a function of the square of the blade length (the circle's radius).

Therefore, a fivefold increase in rotor diameter (from 10 meters on a 25-kW turbine like those built in the 1980s to 50 meters on a 750-kW turbine common today) yields a 55-fold increase in yearly electricity output, partly because the swept area is 25 times larger and partly because the tower height has increased substantially and wind speeds increase with distance from the ground.

Advances in electronic monitoring and controls, blade design and other features have also contributed to a drop in cost.

Optimal configuration of the turbines

Optimal configuration of the turbines to take the best advantage of micro-features on the terrain will also improve a projects productivity.

Cost of financing affects wind energy

Wind energy is capital-intensive, so the cost of financing constitutes a large variable in a wind energy projects economics. For a variety of reasons, financing for wind projects remains more expensive than for mainstream forms of electricity generation.

Project ownership affects cost of financing and the economics of a wind power project. Independent ownership—that is, financing of projects by private power producers on a stand-alone basis, which is how the vast majority of U.S. wind projects are financed—is more expensive than utility-owned financing.

21.4.3   Transmission, tax, environmental

Transmission, tax, environmental and other policies also affect the economics of wind. Transmission and market access constraints can significantly affect the cost of wind energy. Since wind speeds vary, wind plant operators cannot perfectly predict the amount of electricity they will be delivering to transmission lines in a given hour. Deviations from schedule are often penalised without regard to whether they increase or decrease system costs. Interconnection procedures are not standardised and utilities have on occasion imposed such difficult and burdensome requirement on wind plants for connection to transmission lines that wind companies have chosen to build their own lines instead.

As electricity markets are restructured and long-term power purchase agreements give way to trading on power exchanges, transmission and market access conditions will play an increasingly important role in the economics of a wind project.

Federal tax code

The federal tax code, which provides a variety of permanent and temporary incentives for conventional forms of energy, also includes a Production Tax Credit (PTC) for wind energy and a 5-year accelerated depreciation schedule for wind turbines.

The PTC, a key incentive, helps level the economic playing field for wind projects in energy markets where other forms of energy are also subsidized. It must be noted, however, that the current 'on-again, off-again' status of the credit is hobbling project development and the industry as a whole. Uncertainty also affects relationships with vendors and substantially increases costs as orders are rushed to meet PTC deadlines or as planning grinds to a halt and income is lost while the industry awaits an extension. One major U.S. developer stated that a five-year extension of the PTC would provide enough long-term certainty to squeeze an additional 25% out of vendor costs.

Stricter environmental regulations enhance wind energy's competitiveness

Wind powers environmental impact per unit of electricity generated is much lower than that of mainstream forms of electricity generation, as wind energy neither emits pollutants, wastes, or greenhouse gases, nor damages the environment through resource extraction. The higher the air quality and other environmental standards adopted in a country, the more competitive wind energy therefore becomes in the marketplace. Conversely, a relaxation of standards or failure to internalise environmental costs through pollution charges or other processes makes polluting forms of electricity generation appear deceptively cheap. This is an important economic issue, because the hidden 'subsidy' that governments and markets give to polluting energy sources by partially or fully ignoring their health and environmental costs is typically much larger than direct subsidies to such energy sources.

Wind energy and ancillary economic benefits

Wind energy provides ancillary economic benefits:

1. Less dependence on fossil fuels, which can be subject to rapid price fluctuations and supply problems.
2. Steady income for farmers or ranchers who own the land on which wind farms are built and for the communities in which they live (in Texas, for example, ranchers have been reaping income from the wind even as their royalties from oil wells have declined).
3. An increase in the property tax base for rural counties.

Financial and political aspects of wind energy

22.1   Introduction

A financial analysis is intended to clarify whether or not the wind farm project is financially feasible for the chosen site and types of wind turbines and several other parameters. A financial efficiency threshold is often set by the project developers. The financial analysis can reveal, which technical or non-technical modifications can be conducted to reach and surpass the set efficiency threshold. For this purpose a complete cost-benefit analysis under consideration of all relevant project parameters (according to site conditions, choice of technology and financing conditions) must be conducted. There are several major indicators that can be used to evaluate the financial feasibility of a project based on a complete description of benefits and costs. Common examples are:

1. The Financial Internal Rate of Return (FIRR).
2. The Debt Service Coverage Ratio (DSCR).
3. The Return on Equity (ROE).
4. The levelised generation costs of the wind park.

22.2   Costs included in financial analysis

The analysis of costs include:

1. Investment costs.
2. Operation and maintenance costs.
3. Land lease costs.
4. Costs for mitigation measures.
5. Project financing structure.
6. Up-front + administrative costs for Clean Development Mechanism (CDM).

The different types of costs are explained briefly in the following. Certainly the type and number of costs that come up in a project depend on project size and site conditions.

22.2.1   Investment costs

The total investment costs consist of the following items:

1. Turbines including erection.
2. Sea transport and inland transport.

3. Crane, if necessary including sea transport.
4. Crane works.
5. Installation.
6. Civil works.
7. Road access.
8. Crane pads.
9. Foundations/basements.
10. Cable trenches.
11. Control building.
12. Required electrical equipment.
13. Extension of substation.
14. Civil works for new substation.
15. Transformer (132kV/33kV).
16. Auxiliary equipment of substation.
17. 132 kV components.
18. Overhead lines (OHL).
19. Wind park cabling, earthing, SCADA.
20. Distribution stations on site.
21. Electrical equipment inside control building.
22. Auxiliary transformer at control building.
23. Equipment for maintainance team (e.g., tools, cars).
24. Engineering (international and local).
25. Mitigation measures.

If the project has an international character, investment costs should be divided into the costs borne by local investors and the part of costs borne by foreign investors.

22.2.2 Operation and maintenance costs

The O&M costs include:

1. Repairs.
2. Maintenance.
3. Spare parts.
4. Insurance costs.
5. Personnel costs for wind park management and maintenance.
6. Electricity consumption.

The range of the costs is determined by the changing lifetime of the components and their prices which can not be predicted exactly. Furthermore the way of operating the wind parks can have an influence on the expected costs which can also be only estimated for future times. In addition, the size of the machines and the operating time under full load is expected to have an influence on the maintenance costs.

According to consultants experience, manufactures have to guarantee that in wind parks with an installed capacity of approximately 50 MW, two experts (usually, one electrical engineer and one mechanical engineer) will be constantly present at the wind park. Additionally, a team of four local experts has to be established for maintenance tasks and a crane has to be available, at least once per year, to realise operation revisions. Further, condition monitoring should be realised by independent engineers in order to plan the repairs. The O&M costs have taken insurance costs into consideration in the form of an insurance following commissioning (Business Interruption Insurance). This has been calculated as an annual 12.0% of the wind turbine price.

22.2.3 Land lease costs

The agreed land lease costs (with regional authority or private owner of the site) have to be considered.

22.2.4 Costs for mitigation measures

If the analysis of the environmental impacts of the wind park revealed significant problems resulting from erection and operation of the wind park, reasonable mitigation measures have to be planned.

Common environmental problems concern:

1. Protection of birds and bats.
2. Protection of other species disturbed by building and operation of the wind park.
3. Protection of nearby domestic dwellings from noise of the wind park technical resolution measures to prevent adverse effect on telecommunication systems in the area.

The planning of application of mitigation measures is normally controlled during the approval of the planning application of the wind park.

22.2.5 Project financing structure

Depending of the availability of investors and funding, the project developer creates a financing structure: He decides, which part of investments has to be provided as equity and which share is finance based on departments.

1. Equity finance: It has to be considered, which components and working procedures should be financed by equity. Additionally it should be decided whether equity should be used for the first investments.

2. Debt finance: The ultimate financing conditions at which debt can be raised will depend upon several factors, including for example the general environment in the debt markets, local political conditions as well as the Lenders interpretation of the projects risk profile. It has to be decided, whether the debt can be provided by foreign or domestic financial institutions, while local activities will be covered by local loans.

The departments structure describes conditions of repayment and interest by the following parameters:

1. Availability period of the credit.
2. Final maturity: Time period after which repayment should be finalised.
3. Repayment profile: Frequency number of repayment rates is determined
4. Base rate in % per annum.
5. Grace period: A starting period without repayments is often arranged for the first years of project operation.

22.2.6 CDM up-front and administrative costs

In order to register a project as a CDM activity, the project participants have to incur mainly in CDM up-front and CDM monitoring costs. CDM administrative costs are applicable only in years determined as CDM credit periods.

22.3 Expected benefits of a wind project

In case a project has been approved for CDM support, the financial benefits result from electricity sales and sales of Certified Emission Reduction Credits (CERs) of the CDM.

22.3.1 Electricity sales

The returns for generated electricity depend on the political framework conditions (and the related support mechanisms) applied in the country. In several countries feed-in tariffs are paid for the delivery of each kWh to support the development of renewable energy technology. If there is no support mechanism, returns of electricity generation are calculated using the prevailing tariffs for electricity sale as well as forecasts of the development of these tariffs.

22.3.2 CDM revenues

The CDM scenario includes the additional Certified Emission Reduction (CER) revenues as cash inflows, as well as the CDM up-front costs and administrative costs as cash outflows. To calculate the revenues from CER sales, the amount of CO_2 avoided must be determined. This is done by CDM baseline-assessment. The expected amount of CERs, which can be saled is multiplied by the assumed price of CERs for the crediting period.

22.4 Results of a financial analysis - Major financial indicators

If costs and benefits of the project are determined, major financial indicators may be calculated. To take into account financial process like depreciation and inflation several additional country-specific values must be determined:

1. Inflation rate.
2. Rate of exchange (if foreign investment and project support is part of the project).
3. Depreciation rates.
4. Dividend distribution: The mode of dividend calculation for investors have to be determined. As an example dividends can be geared to internal funds available for distribution (cash after debt service, taxes and reserve account payments) and net income. Dividend distribution may be limited to the case that the Debt Service Coverage Ration (DSCR) exceeds a certain factor and net income is positive.
5. Corporate taxes: Country-specific information about corporate taxes has to be collected to define the fees, which have to be paid by the wind park operator.
6. Import taxes: Country-specific import taxes for the wind park components have to be found out.
7. Discount rate Weighted Average Cost of Capital (WACC).

22.4.1 Discount rate weighted average cost of capital (WACC)

The discount rate to be normally used in financial benefit-cost analyses is the WACC. The WACC represents the cost incurred by the entity in raising the capital necessary to implement the project. Since most projects use several sources to raise capital and each of these sources may seek a different return, the WACC represents a weighted average of the different returns paid to these sources.

The methodology used to calculate the WACC is as follows:

Step 1: A categorisation of financing components according to the Project Financing Plan is done. These components can be divided into domestic and foreign components. The two types of components often have different loan conditions (e.g., local loan at an interest rate of 7% p.a. and local equity share/ foreign loan at an interest rate of5 % and international equity).

Step 2: The cost of funds are estimated. Since government funds are not costless they might be applied to purposes other than the project, such as debt repayment or to alternative investments. For simplicity, the average cost of government funds can be calculated by dividing total government debt servicing by total public debt. In order to estimate the cost of equity capital, the Capital Asset Pricing Model (CAPM) can be used (Fig. 22.1). CAPM describes the relationship between risk and expected return and that is used in the pricing of risky securities.

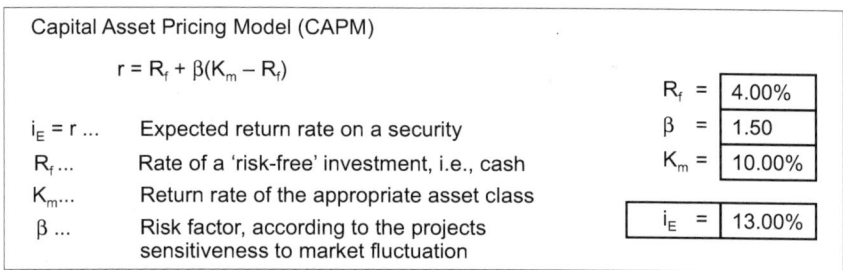

Figure 22.1: Capital asset pricing model.

The general idea of CAPM is that investors need to be compensated in two ways – time value of money and risk. The time value of money is represented by the risk-free (R_f) rate in the formula and compensates the investors for placing money in any investment over a period of time. The other half of the formula represents risk and calculates the amount of compensation the investor needs for taking on additional risk. This is calculated by taking a risk measure (β) that compares the returns of the asset to the market over a period of time and to the market premium (K_m-R_f).

Step 3: Adjustment for taxation. If interest payments are deductible for taxation, applicable tax rates have to be adjusted to each component.

Step 4: Adjustment for domestic inflation. The estimated costs of borrowing and equity capital have been adjusted for inflation to obtain the WACC in real terms. Domestic inflation rate (of 2 %) has been used for domestic loans and equity. Shadow price adjustment does not apply in the financial analysis and hence, the Standard Conversion Factor (SCF) of 1 has been considered for local currency expenditure, without reducing local costs.

Step 5: Application of the minimum rate test, i.e., to review the real cost of capital for each component. The rate for each component should be at least 4%.

Step 6: The determination of the WACC has been done by applying the weighting percentage to each component.

22.4.2 Major financial indicators

After all necessary parameters are defined, the financial internal rate of revenue can be calculated using the feed-in tariff or selling price for energy to forecast returns of the wind project. All values are integrated into a financial model, which calculates the FIRR for the assumed scenarios. However it is possible as well to calculate the necessary selling price (or feed-in tariff) to reach a certain amount of revenue on equity. Financial indicators are used to evaluate projects and to allow comparison between different sites and conditions. Major financial indicators are discussed below.

Net present value

The Net Present Value (NPV) of an investment has been defined as the present (discounted) value of future cash inflows minus the present value of the investment and any associated future cash outflows. A positive NPV indicates that the projects are justified in an economic sense and vice verse for a negative NPV.

Financial IRR (FIRR)

The Financial Internal Rate of Return (FIRR) is an indicator to measure the financial return on investment of an income generation project and is used to make the investment decision. The FIRR is obtained by equating the present value of investment costs (as cash outflows) and the present value of net incomes (as cash in-flows) as shown in Fig. 22.2.

$$I_0 + \frac{I_1}{(1+r)^1} + \frac{I_2}{(1+r)^2} + \cdots + \frac{I_m}{(1+r)^m} + \cdots + \frac{..B_1}{(1+r)^1} + \frac{..B_2}{(1+r)^2} + \cdots + \frac{..B_m}{(1+r)^m} + \cdots$$

Figure 22.2: FIRR ROE calculation.

with:

I_x = Initial investment
B_x = Benefits
r = discount rate
m = period

Return on equity (ROE)

The ROE measures the profitability of a project, calculated as net income divided by shareholders equity. Essentially, ROE reveals how much profit a project generates for the capital shareholders, which have invested in it. At least it turns out that a project revenue cannot grow faster than its current ROE without raising additional cash.

Debt service coverage ratio (DSCR)

The Debt Service Coverage Ratio (DSCR) defines the capability of the Wind Farm operating company to repay principal and interest payments on debt. It is stated as the ratio between cash available to service debt and total debt service.

Levelised costs

Similar to the internal rate of return calculation, the levelised costs of energy are calculated by searching for a tariff for electricity, with which the net present value turns out as zero and the internal rate of return equals the applied discount rate. In this calculation only the annual project costs such as operation and maintenance and debt service are included, whereas the inflation rate is set to zero.

22.5 Sensitivity analysis of project parameters

The purpose of the sensitivity testing is to establish which project parameters have the potential to alter the financial feasibility of the project by the greatest amount. By systematically altering individual parameters and recording the influence on the project evaluation criteria, those parameters with the greatest potential to cause the economics of the project to deviate from its expected value can be isolated. This provides a transparent overview of the projects risk profile, assists project planning and helps prepare risk mitigation strategies.

22.5.1 Methodology

Where applicable, expected values of parameters of the financial analysis are (as used in the base case with CDM calculations) increased and decreased by +5%, –5% and +10%, –10% respectively. This is done for one parameter at a time, so as to isolate the impact of that variable on project feasibility. Complimentary to this, the respective Sensitivity Indicators (SI) and Switching Values (SV) can be derived.

The definitions of the SI and SV are as follows:
1. SI: represents the percentage change in project NPV as a result of a 1 % increase in the parameter.
2. SV: represents the percentage change in the parameter value necessary to drive project NPV down to zero.

Sensitivity variables can be for example:
1. Investment costs.
2. Energy generation.
3. Routine O&M.
4. Sales tariff.
5. CO_2 contract price.

22.6 Political aspects of wind energy

Country-specific conditions for the development of wind energy projects are based upon the policy framework conditions prepared by national or regional governments. The financial feasibility of any wind energy project proposed to sell electricity to the grid depends on the available framework conditions for support. Inadequate or non-existent framework conditions often form crucial barriers impeding the exploitation of available wind energy potential. Despite macroeconomic benefits which can arise when the supply situation and environmental factors are taken into account, it can happen that a large number of investment projects are not translated into reality.

In many countries accompanying government measures supporting the development of wind energy do not exist. Without any security about selling prices, feed-in tariffs, possible tax rebates or assistance for capacity building, it is hard to create sufficient security for large investments as required for the development of a wind park.

22.7 Strategies and support mechanisms for wind energy

There are different types of instruments available to support renewable energy projects and these range from direct to indirect policy instruments. Direct policy measures such as a Renewable Energy Feed-in Tariff (REFIT) aim to stimulate the installation of renewable technologies immediately, whereas indirect instruments focus on improving long-term framework conditions. Besides regulatory instruments, there are other approaches for promoting and supporting renewable energy, primarily aimed at removing the barriers which limit investments in renewable energy such as lack of information, lack of skills, limited research and development, inadequate regulatory structures and limited incentive programmes. The different types of strategies and supporting mechanisms are detailed below.

22.7.1 Incentives of renewable energy

Renewable energy incentive programmes assist in establishing a competitive self-sustaining renewable energy supply while increasing the quantity of renewable energy generated countrywide. There are a range of incentives that can be used such as:

1. Direct subsidies from the government in the form of capital grants.
2. Fiscal incentives such as tax rebates for project developers and consumers or tax exemptions for the import of renewable energy technology/ equipment.
3. Investment incentives which encourage the participation of national and international financiers.
4. Incentives that promote public-private partnerships to increase the use of renewable energy technologies.
5. Power production incentives.

22.7.2   Institutional support

Clear, long-term legislation and policies that support renewable energy have a critical role to play in building investor confidence and in ensuring the sustainable growth of the renewable energy sector. A renewable energy bill for instance, would be designed to complement a broader suite of initiatives such as existing government programmes and other new laws that might be put in place such a green procurement law (which could include renewable energy use as one of the criteria). Other forms of institutional support can come in the form of infrastructure and planning regulations such as building codes which could treat renewable energy projects as 'privileged' or 'special' projects where local authorities are required to designate specific priority or preferential zones for renewable energy utilisation.

22.7.3   Information portals

There is often a lack of clarity by both developers and also key institutions as to how to get projects moving, especially with regard to the many legal and regulatory requirements at both a local and national level. In order to encourage, support and facilitate the development of renewable energy projects, in many countries institutions have been established at both local and regional levels to provide a variety of functions, from providing basic information to developers, through to actively engaging and assisting developers in submitting projects to the relevant authorities, or even processing applications.

Such an institutional entity, often referred to as a one-stop shop, is aimed at providing the necessary information and support required by developers and financiers. The one-stop shop could either exist simply as an information portal or it could actually take a proactive approach to assist developers navigate the complicated licensing and permit requirements. Probably these services and functions may not be required by all developers, especially the larger developers who would have much greater internal infrastructures to develop projects, however the aim is to provide a range of services and information to a wide range of clients based on the needs of those clients.

In some cases one-stop-shops have been incorporated into existing organisations, in others dedicated organisations have been established. The form of these agencies differs by location. In most instances they are national energy agencies nested within government ministries, in others they are located within local economic development agencies.

In the Mediterranean a number of countries have formed a regional body, called the Mediterranean Renewable Energy Centre (MEREC) which is geared not only towards investors but aims to develop regional competencies through the dissemination of information, training of staff and technology transfers. In the United States a more general national body oversees developments within the field but, on a lower level individual states run one-stop shops or independent agencies which among other activities market local investment opportunities in renewable energies.

22.7.4 Industry development

Industry development is an important aspect for both regional and national growth. It is important for provincial governments to do what they can to help promote a successful industry covering everything from actual raw materials to developing and financing projects to consulting and other ancillary services associated with a thriving regional renewable energy market. A proactive approach to renewable industry development will not only create jobs and revenues, but it will also help in long-term development of the region and importantly assist in energy security and independence.

Incorporating industry development is an integral component to a successful regulatory action plan to help foster the growth of the renewable energy market and help the region achieve its goals of clean energy capacity. There are many aspects to industry development beyond actions such as tax incentives/relief and relaxing licensing requirements to attract businesses.

As part of a comprehensive industry development plan it is important to assist and incentivise the establishment of industry or trade associations. Industry associations or trade groups can help bring a unanimous voice to renewable energy related businesses in the region, which can help policymakers identify pros and cons of regulatory legislation and identify what steps need to be taken in the future to help reform or develop the market. In addition, trade associations can help with shaping public opinion to assist in the market growth.

Also, of importance to stimulate the industry is research and development. It is an important task for energy related ministries to support local university and institutes through research grants, funding of research projects and channelling public funding to secure Intellectual Property (IP) rights. Government could help develop or actively participate in both provincial and national level research advisory committees working on the identification of important research needs of the different regions of a country.

One of the most direct supporting mechanisms for renewable energy industry development is training and capacity building. Capacity building could be done through trade associations, workshops, training programmes, or partner with international and non-governmental organisations that assist countries with these types of activities.

22.8 Environmental impact assessment of wind energy

An additional support mechanism to help facilitate the development of renewable energy projects is a platform where developers and financiers could be made aware of each other. Situations in the market do arise when financing institutions or agencies are looking to invest in renewable energy project developers that have identified project opportunities and successfully completed a number of prerequisites, e.g., site location, REFIT approval and/or Environmental Impact Assessment (EIA). Likewise, there are also project developers who have identified renewable energy project opportunities and require some level of structured finance.

A platform where financiers and developers could exchange information about their needs and opportunities could prove to be invaluable in facilitating the development of renewable energy projects and the market. In addition, such a platform could host information regarding international organisations, NGOs and/or charities that are looking to finance clean energy projects or particular steps, such as EIAs, stakeholder consultations, feasibility study, etc. A platform such as this could be hosted by the renewable energy project one-stop shop. A project developer and financier forum could be hosted on the website that will be developed for the one-stop shop. It could have the contact details of all those developers that have identified renewable energy opportunities, as well as a list of local, national and international companies, NGOs and organisations that are willing to fund projects.

22.9   International examples

22.9.1   United States – Texas

In the State of Texas the renewable energy sector (primarily wind) has grown over the past two years with over 3900 MW of installed wind generated electricity. In 2013 alone there was a record of a Gigawatt of wind electricity installations. The Department of Energy (DOE) reported wind to be the fastest growing renewable energy technology, which had grown by approximately 45%. The growth was due to a strong demand, investment of private capital and support from federal and state governments using various incentives.

Incentives are seen as a means of supporting renewable energy technological developments and to help reduce the up-front capital expenditure of investing in renewable energy systems. There are a variety of incentives which reduce costs by means of tax exemptions, funding for project implementation and feed-in-tariffs. The goal is to make wind energy more cost-competitive against traditional fossil fuelled generated electricity. In Texas there is a franchise tax exemption for manufacturers, sellers and installers of renewable energy systems. A franchise tax is a tax levied by some U.S. states for corporations, which is based on the number of shares that are issued or in some instances, the amount of their assets. Business owners can deduct the total cost of the system from companies' taxable income. Texas also has 100% property tax exemption for the appraised value of the site which has a wind and solar energy generating system. These tax exemptions are all part of the Texas Tax Code. Another form of tax incentive is the Federal Renewable Electricity Production Tax Credit. Production Tax Credits (PTCs) allow companies that invest in renewable energy generation to write off or deduct their investment against other investments they make. The tax deductions from the PTC allows for 2.1 credit per kWh against corporate income tax for electricity generated during the first 10 years of wind power operation. This PTC is regarded as an important incentive for wind growth as it encourages private investment in wind energy projects.

Another example of an incentive provided by federal and state governments would be grants, such as the United States Department of Agriculture (USDA) renewable energy and efficiency programme. This programme provides grants and loan guarantees for use by farmers, ranchers and small rural businesses that invest in renewable energy and energy efficiency projects in rural areas.

22.9.2   Germany

In 2013, renewable energy in Germany rose by 16% of the total electricity generated. These figures saw a turnover of 16.2 billion from the erection of clean power plants and 15.2 billion from the operation of the plants, as well as creating approximately 214,000 jobs and preventing 101 MT of CO_2 emissions.

The wind sector in Germany has developed to become a world leader – Germany has more electricity generated by wind than any other country in the world. The first incentive seen in Germany promoting wind generation was started in 1989 by an incentive known as '100 MW of wind' created by the Ministry of Research and Technology. Under this programme wind projects were paid a subsidy of 0.04 per kWh generated. In addition any project that was given a subsidy had to disclose the turbine and wind farms performance. This coupled financial incentives with a research project that would help to lay the foundation of future technological developments and planning policies. This incentive proved popular and a year after its inception the quota was increased to 250 MW. A small-scale feed-in-tariff was introduced as part of the Electricity Grid Feed Act, which saw producers receiving 0.0849 cents per generated kWh and focused mainly on coastal areas with turbines ranging in size from 20 to 150 kW output.

The impressive take-up rate of wind energy in Germany is due to strong support mechanisms, such as the Renewable Energy Sources Act (EEG). The EEG was implemented in 2000 to replace the previous Electricity Grid Feed Act. The core element of the EEG legislation was to impose a priority purchase obligation, i.e., the grid operators were obligated to connect renewable energy producers regardless if there are utilities, businesses or private residential households to the grid. This was done so that renewable energy became the priority source of electricity above any other source. Potential consequences of this Act might be that conventional fossil fuelled power sources would have to reduce their generation to accommodate for clean energy being fed into the grid. This also saw increased investor security in renewable energy and ensured that every unit of renewable energy would be sold. Germany has focused on installing some of the most powerful wind turbines in a given area as possible. Due to improvements in technological development of wind turbine design, which includes larger and more efficient wind turbines, Germany is able to maximise each site's potential for installed capacity.

Another government support mechanism took the form of a Market Incentive Programme for renewable energy (MAP) which was implemented in 1999. Intended to support solar power generation for the residential homes market the Federal Office of Economics and Export Control (BAFA) incentivised investors both private individuals and small-to-medium sized enterprises with grants for investments in solar thermal energy projects. Larger enterprises were eligible for reduced interest rate loans that were available specifically for renewable energy systems.

These various incentive mechanisms helped Germany achieve more than 15,000 MW of installed wind power capacity as early as 2009. In addition by 2012, the wind energy sector alone contributed to 9.64 billion in revenues to the market total of 17.4 billion. Also the wind energy market was employing 73,800 people out of a total of 214,000 employed in the German clean energy market. It is quite clear that with good market support mechanisms and strategies Germany was able to become a world leader in the wind energy market in a relatively short period of time.

AWEA	American Wind Energy Association
Bloomberg NEF	Bloomberg New Energy Finance
BPA	Bonneville Power Administration
BOEM	Bureau of Ocean Energy Management
CAISO	California Independent System Operator
CREZ	Competitive Renewable Energy Zone
DOE	U.S. Department of Energy
EDPR	EDP Renováveis
EEI	Edison Electric Institute
EIA	U.S. Energy Information Administration
ERCOT	Electric Reliability Council of Texas
FERC	Federal Energy Regulatory Commission
GE	General Electric Corporation
GW	Gigawatt
HTS	Harmonized Tariff Schedule
ICE	Intercontinental Exchange
IOU	Investor-owned utility
IPP	Independent power producer
ISO	Independent system operator
ISO-NE	New England Independent System Operator
ITC	Investment tax credit
kV	Kilovolt
kW	Kilowatt
kWh	Kilowatt-hour
m^2	Square meter
MISO	Midcontinent Independent System Operator
MW	Megawatt
MWh	Megawatt-hour
NERC	North American Electric Reliability Corporation
NREL	National Renewable Energy Laboratory

NYISO	New York Independent System Operator
O&M	Operations and maintenance
OEM	Original equipment manufacturer
PJM	PJM Interconnection
POU	Publicly owned utility
zPPA	Power purchase agreement
PTC	Production tax credit
REC	Renewable energy certificate
RGGI	Regional Greenhouse Gas Initiative
RPS	Renewables portfolio standard
RTO	Regional transmission organization
SPP	Southwest Power Pool
USITC	U.S. International Trade Commission
W	Watt
WAPA	Western Area Power Administration

Alternating current (AC): An electrical current that reverses direction at regular intervals or cycles. In the United States, the standard is 120 reversals or 60 cycles per second. Electrical grids in most of the world use AC power because the voltage can be controlled with relative ease, allowing electricity to be transmitted long distances at high voltage and then reduced for use in homes.

Anemometer: Measures the wind speed and transmits wind speed data to the controller.

Bird mortality: Mortality from bird collisions with the turbine blades, towers, power lines, or with other related structures and electrocution on power lines.

Blades: The aerodynamic surface that catches the wind. Most commercial turbines have three blades.

Braking system: A device to slow a wind turbine's shaft speed down to safe levels electrically or mechanically.

Capacity factor: The average power output of a wind development divided by its maximum power capability, its rated capacity. Capacity factor depends on the quality of the wind at the turbine. Higher capacity factors imply more energy generation. On land, capacity factors range from 0.25 (reasonable) to over 0.40 (excellent). Offshore, capacity factors can exceed 0.50.

Carbon dioxide (CO_2): A naturally occurring gas and also a by-product of burning fossil fuels and biomass as well as landuse changes and other industrial processes. CO_2 is the principal greenhouse gas that is produced by human activity and influences climate change.

Climate change: Changes in a climate system over decades or longer. The term often refers to changes in climate that can be attributed directly or indirectly to human activities that altered the composition of the global atmosphere – changes that are beyond the natural climate variability observed over comparable time periods.

Controller: The controller starts up the turbine generator at wind speed of about 8 to 16 mph and shuts off the generator at about 65 mph.

Cut-in speed: The wind speed at which the turbine blades begin to rotate and produce electricity, typically around 10 mph.

Cut-out speed: The wind speed at which the turbine automatically stops the blades from turning and rotates out of the wind to avoid damage to the turbine, usually around 55 to 65 mph.

Direct current (DC): A type of electrical current that flows only in one direction through a circuit, usually at relatively low voltage and high current. To be used for typical 120 or 220 volt household appliances, DC must be converted to AC, its opposite. Most batteries, solar cells and turbines initially produce direct current which is transformed to AC for transmission and use in homes and businesses.

Direct employment: The total number of people employed in companies belonging to a specific sector.

Efficiency: Describes the amount of active electrical power generated as a percentage of the wind power received by the turbine.

Emissions: The discharges of pollutants into the atmosphere from stationary sources such as smokestacks, other vents, or the surfaces of commercial or industrial facilities and mobile sources such as motor vehicles, locomotives and aircraft. With respect to climate change, emissions refer to the release of greenhouse gases into the atmosphere over a specified area and period of time.

Employment: Indirect employment includes employment throughout the production chain, including those who supply raw materials and intermediate components to, for example, a wind turbine.

Energy payback: The time period it takes for a wind turbine to generate as much energy as is required to produce the turbine, install it, maintain it throughout its lifetime and, finally, scrap it.

External costs: Costs incurred in activities which may cause damage to a wide range of receptors, including human health, natural ecosystems and the built environment and yet are not reflected in the monetary price paid by consumers.

Gear box: Gears connect the low-speed shaft to the high-speed shaft and increase the rotational speed of the shaft to the speed required by the generator. The gear box is heavy and power losses from friction are inherent in any gearing system.

Generator: A device that produces electricity from mechanical energy, such as from a rotating turbine shaft.

Geographic information system (GIS): A software system which stores and processes data on a geographical or spatial basis. GIS can be used to evaluate potential sites for wind farms and consider a variety of geographic data simultaneously, such as wind conditions, population and bird migration routes.

Grid-connected system: A residential electrical system such as solar panels or wind turbines which act like a central generating plant, supplying power to the commercial grid.

Guy wires: A tensioned cable designed to add stability to tall, narrow structures, frequently used to support ship masts, radio masts and wind turbines.

Inverter: A device that converts direct current (DC) electricity to alternating current (AC) either for standalone systems or to supply power to an electricity grid.

Life cycle assessment: An evaluation of the environmental impacts of a given product or service, such as a wind farm, throughout its life cycle, including manufacturing the parts, installation, operation and disposal.

Load: In reference to wind developments, load describes the energy demand placed on a power generating facility or the energy consumed by a group of customers or set of equipment. Usually expressed in amperes or watts.

Megawatt (MW): 1000 kilowatts (kW) or 1 million watts (W), standard measure of electric power generating capacity. Large utility scale wind turbines usually produce 900 kW to 2 MW per turbine.

Megawatt-hour (MWh): The amount of energy used if work is done at an average rate of 1 million watts for 1 hour.

Nacelle: The nacelle sits atop the tower and contains the gearbox, shafts and generator of a wind turbine. Some nacelles are large enough for a helicopter to land on.

Nitrogen oxides (NO$_x$): Refers to NO and NO$_2$, which are often produced by combustion under high pressure and high temperature, such as in the engine of a car. NO$_x$ is a major contributor to acid depositions, the formation of ground-level ozone in the troposphere and photochemical smog.

Offshore wind developments: Wind projects installed in shallow waters off the coast. Turbine construction has to be modified to accommodate the depth of the water. A number of countries have built offshore wind developments, but currently there are no offshore wind farms in the U.S.

Onshore wind developments: Wind farms installed on land. Land-based wind farms are significantly cheaper to build than offshore facilities, but the wind is generally stronger and more steady offshore.

Participatory planning: A planning process open to high levels of public engagement. The success of a wind farm development is influenced by the nature of the planning and development process and public support tends to increase when the process is open and participatory. Thus, collaborative approaches to decision-making in wind power implementation can be more effective than topdown, imposed decision-making.

Pitch: The angle between the edge of the blade and the plane of the blade's rotation. Blades are turned, or pitched, out of the wind to control the rotor speed.

Rated wind speed: The wind speed at which the turbine is producing power at its rated capacity. The rated wind speed generally corresponds to the point at which the turbine can perform most efficiently. Because of the variability of the wind, the amount of energy a wind turbine actually produces is lower than its rated capacity over a period of time.

Renewable energy: Energy which comes from renewable resources such as sunlight, wind, rain, tides and geothermal heat, which are naturally replenished. Fossil fuels, such as coal and oil, are considered non-renewable resources because they are consumed much faster than nature can create them.

Renewable portfolio standards (RPS): Laws developed federally or by states mandating that electricity providers obtain a minimum fraction of their energy from renewable resources.

Rotor hub: The center of a turbine rotor, which holds the blades in place and attaches to the shaft. The rotor refers to both the turbine blades and the hub.

Shaft: The rotating part in the center of a wind turbine or motor that transfers power. A high-speed shaft drives the generator. A low-speed shaft is turned by a rotor at about 30 to 60 rpm.

Storage battery: Also known as a rechargeable battery. Can transform energy from electric to chemical form and *vice versa.* Can be connected to turbines to temporarily store electricity in chemical form during peak production times and release electrical energy when demand is high.

Substation: A facility that steps up or steps down the voltage in utility power lines. Voltage is stepped up where power is sent through long-distance transmission lines. It is stepped down where the power is to enter local distribution lines.

Tower: The base structure that supports and elevates a wind turbine rotor and nacelle.

Transformer: Multiple individual coils of wire wound on a laminate core. Transfers power from one circuit to another using magnetic induction, usually to step voltage up or down. Works only with AC.

Wind lease: An agreement signed by a landowner that grants a developer the right to use their land for wind development and in return, provides compensation to the landowner. Typically, the developer owns any turbines that are put up and does all of the work of developing the project. Wind leases are binding legal documents that typically cover 30 to 60 years or more. These agreements can allow turbines to be constructed on privately owned, actively farmed land.

Wind power class: A way of quantifying on a scale the strength of the wind at a project site. The department of energy defines the wind class at a site on a scale from 1 to 7 (1 being low and 7 being high) based on average wind speed and power potential to offer guidance about where wind projects might be feasible.

Wind turbine: A machine that captures the force of the wind. Also called a wind generator: when used to produce electricity. Most commercial wind generators are horizontal axis wind turbines. If wind energy is used directly by machinery, such as for pumping water, cutting lumber or grinding stones, the machine is called a windmill.

Wind vane: Measures wind direction and communicates with the yaw drive to orient the turbine properly with respect to the wind.

Yaw: To rotate around a vertical axis, such a turbine tower. The yaw drive is used to keep a turbine rotor facing into the wind as the wind direction changes.

References

Allen, F.K., *Principles of Energy Conversion*, Mcgraw-Hill, New York.

Andrew, Swift, *Wind Energy Essentials*, Wiley, New York.

Anthony, L.R., *Wind Energy*, McGraw-Hill, New York.

Bianchi, F.D., Wind Turbine Control Systems, Interscience, New York, USA.

Billmeyer, F.W., *Renewable Energy Resources,* Reinhold, New York.

Faulk, B.F., *Wind Power Energy*, Van Nostrand Reinhold, New York.

Gacher, R.M., *Environmental Aspects of Renewable Energy Sources*, Elsevier Applied Science, London.

Hau, Erich, *Wind Turbines*, Springer-Verlag Berlin Heidelberg.

Hernán, D.B., *Renewable Energy Sources and Emerging Technologies*, Van Nostrand Reinhold, New York.

James, F.M., *Essentials of Wind Energy*, Wiley, New York.

Jon, G. McGowan, *Wind Energy Basics*, Chelsea Green Publishing, USA.

Julia, K., Johann, *Offshore Wind Energy*, Springer-Verlag Berlin Heidelberg.

Lewis, F.W., *Commissioning of Wind Projects*, Henser, USA.

Martin, J. Pasqualetti, *Wind Power*, John Willey-Interscience, New York.

Paul, Gipe, *Handbook of Wind Energy*, Chelsea Green Publishing, USA.

Peter, Barth, *Wind Energy*, Springer-Verlag Berlin Heidelberg.

Ricardo, M.J., *Fundamentals of Renewable Energy Processes*, Academic Press

Richard, P. Walker, *Wind Power and Power Politics*, Routledge, UK.

Rogers, D.F., *Non-conventional Energy Sources and Utilisation*, Noyes, UK.

Siegfried, Heier, *Grid Integration of Wind Energy Conversion Systems*, Wiley, New York.

Thomas, Ackermann, *Wind Power in Power Systems*, Wiley, New York.

Tore, Wizelius, *Developing Wind Power Projects*, Earthscan, UK.

Vaughn, Nelson, *Wind Energy: Renewable Energy and the Environment*, CRC Press, New York.

Index